Lubrication Degradation

Reliability, Maintenance, and Safety Engineering: A Practical Field View on Getting Work Done Effectively

Series Editor: Robert J. Latino, Reliability Center, Inc., VA

This series will focus on the "been there, done that" concept in order to provide readers with experiences and related trade-off decisions that those in the field have to make daily, between production processes and costs no matter what the policy or procedure states. The books in this new series will offer tips and tricks from the field to help others navigate their work in the areas of Reliability, Maintenance and Safety. The concept of 'Work as Imagined' and 'Work as Done', as coined by Dr. Erik Hollnagel (an author of ours), this series will bridge the gab between the two perspectives to focus on books written by authors who work on the front-lines and provide trade-off decisions that those in the field have to make daily, between production pressures and costs...no matter what the policy or procedure states. The topics covered will include Root Cause Analysis (RCA), Reliability, Maintenance, Safety, Digital Transformation, Asset Management, Asset Performance Management, Predictive Analytics, Artificial Intelligent (AI), Industrial Internet of Things (IIoT), and Machine Learning (ML).

Lubrication Degradation
Getting into the Root Causes
Sanya Mathura and Robert J. Latino

For more information on this series, please
visit: https://www.routledge.com/Reliability-Maintenance-and-Safety-Engineering-
A-Practical-Field-View-on-Getting-Work-Done-Effectively/book-series/
CRCRMSEGWDE

Lubrication Degradation
Getting into the Root Causes

Sanya Mathura and Robert J. Latino

CRC Press
Taylor & Francis Group
Boca Raton London New York

CRC Press is an imprint of the
Taylor & Francis Group, an **informa** business

First edition published 2022
by CRC Press
6000 Broken Sound Parkway NW, Suite 300, Boca Raton, FL 33487-2742

and by CRC Press
2 Park Square, Milton Park, Abingdon, Oxon, OX14 4RN

Library of Congress Cataloging-in-Publication Data

Names: Mathura, Sanya, author. | Latino, Robert J., author.
Title: Lubrication degradation : getting into the root causes / Sanya
Mathura, Robert J. Latino.
Description: First edition. | Boca Raton, FL : CRC Press, 2022. | Series:
Reliability, maintenance, and safety engineering: a practical field view
| Includes index.
Identifiers: LCCN 2021042073 (print) | LCCN 2021042074 (ebook) | ISBN
9781032171579 (hbk) | ISBN 9781032171586 (pbk) | ISBN 9781003252030
(ebk)
Subjects: LCSH: Lubrication and lubricants--Deterioration.
Classification: LCC TJ1077 .M3599 2022 (print) | LCC TJ1077 (ebook) | DDC
621.8/9--dc23/eng/20211027
LC record available at https://lccn.loc.gov/2021042073
LC ebook record available at https://lccn.loc.gov/2021042074

ISBN: 978-1-032-17157-9 (hbk)
ISBN: 978-1-032-17158-6 (pbk)
ISBN: 978-1-003-25203-0 (ebk)

DOI: 10.1201/9781003252030

Typeset in Times
by Deanta Global Publishing Services, Chennai, India

Contents

Foreword

At a very early age, we are taught that failure is unacceptable. From the time you brought home an "F" on your math test to when you let a goal score during your soccer game, failure was an unacceptable result stemming from a lack of performance or worse – inattentiveness, laziness, or stupidity. Failure can find its root cause in lackluster performance due to many reasons but it doesn't have to be the villain we have accepted it to be. Failure is a key to success if you are willing to use it to open the door of opportunity. You have to be willing to take a leap of faith and trust that you can use what *failure* can offer. There are countless examples of people who have turned failure into success. Look up the history of Silly Putty®, Teflon®, the microwave oven, and Post-it Notes®, and you will see that each of these items found its genesis in failure. The same can be said of notable people such as Michael Jordan, Oprah Winfrey, Dr. Seuss (Theodor Geisel), and even Milton S. Hershey (the Chocolate guy); each became exceptionally successful due, in part, to how they reacted to failure. Failure was (and is) an opportunity.

Mathura and Latino have successfully developed a work that examines the failure of lubricating oil. Their intent is to look beyond the physical roots assigned to the various lubrication degradation mechanisms and dive deeper into the human and systemic roots that are often overlooked. The thesis is that these issues are recurring since the root cause of the issue was never rectified in the first place. The book is predicated on the concept of applied Root Cause Analysis (RCA) for failure mechanisms such as oxidation, thermal degradation, microdieseling, electrostatic discharge, additive depletion, and contamination.

Those who design and blend lubricants will find this work to be a valued reference. Those who market and sell lubricants will use this work as a means to understand performance in order to parley it into their marketing collateral and value statements. The end user of lubricants – be it an original equipment-manufacturer, maintenance/reliability department, or condition-monitoring laboratory – will use this for increased insight and remedies.

When failure occurs, we must look deeper into the reasons and rationale. Dropping our heads, sulking, and accepting that it is the way it is, and nothing can be done is a defeatist attitude. Taking such a stance will only lead to stifled advancement with little hope for a positive outcome. Only by deeply analyzing the root cause of the failure and then attempting (and it will be an attempt because there are no guarantees for success) to change the state in order to thwart, if not eliminate, the failure, will success and increased performance prevail. In order for this to happen, you will need insight. That insight finds its foundation in information. Information can be transformed into knowledge, which will morph into wisdom only if the knowledge is applied and the analysis continues. There is a long row to hoe, yet using this work from Mathura and Latino, the ground is loosened and ready for seed.

Michael D. Holloway, President of 5th Order Industry

Preface

First off, thanks for your interest in lubrication degradation mechanisms (LDMs) and Root Cause Analysis (RCA). It is certainly a niche combination of fields, but they are interdependent when it comes to each other's success.

Our intent with this book was to combine Sanya's deep subject matter expertise in LDM with Bob's deep domain knowledge about the process of holistic RCA.

This is written as a Focus Book, so it is more geared toward a "how-to, no fluff" book that gets to the point quickly. The emphasis is on practicality versus theory and is written to allow anyone within either of the fields to simply pick the book up and begin applying the concepts.

The book first introduces the reader to the basic concepts involved in lubrication. We thought this element to be critical as the basics are sometimes misunderstood, which can lead to misconceptions in the future. The reader is also introduced to the different types of wear mechanisms so these can be applied later.

Analysts are also provided a comprehensive understanding of what a logic tree is, how to properly construct one, and how to use it to tell a "story" to their managers about the "hows" and "whys" that triggered a lubrication-related failure.

We also explore the "root system" of failures in general, which consists of:

1. Physical Root Causes (component-level causes)
2. Human Root Causes (active decision-making causes)
3. Latent or Systemic Root Causes (organizational system causes)

We wrote this book with the intention of it being a field reference guide for those in the front lines who deal with the six most common lubrication degradation mechanisms. They are:

1. Oxidation
2. Microdieseling
3. Additive Depletion
4. Electrostatic Spark Discharge (ESD)
5. Contamination
6. Thermal Degradation

Each of these mechanisms is defined as how to identify and properly classify them; then we delve into using a logic tree to explore all of the most common hypotheses that can cause these mechanisms to surface. This essentially provides our readers an *LDM troubleshooting flow diagram*, which they can take to the field and kick-start their RCA immediately.

Each logic tree comes complete with a comprehensive verification log (VL). The VL suggests verification methods that the analyst can use to validate whether certain hypotheses are "True" or "Not True". The rest is simple as the evidence guides the analysis, not the analyst.

This is where the physical sciences and the social sciences come together and depend on each other, to ensure a successful RCA. We go to great lengths to demonstrate why blaming a decision-maker is counterproductive (and even dangerous). We are not interested in "who" made an inappropriate decision, but more interested in "why" they thought it was the right decision at the time. It is this reasoning that separates a Root Cause Analysis from a Shallow Cause Analysis.

We then wrap up the book by bringing all of the learnings from the previous chapters into a single, practical case study. Unlike the other chapters that explored how those failure mechanisms could occur in general, the last chapter actually deals with a specific case where a certain LDM did occur, and the analyst uses the logic tree approach to walk through the complete RCA – thus practicing what we have been preaching!

Will everybody who reads this text agree with its content? No. Can they benefit regardless? Yes. We hope to spark debate within the minds of our readers where the differences are contrasted between how we approach LDM and RCA and how they are currently conducting them at their facilities.

Perhaps we will sway some to agree with certain premises in this text, and others will improve upon their current approaches with the ideas presented. Either way, the journey of learning is what is most important. Analysts will collect the necessary data, sift out the facts, and make their own determination as to what they believe is best for them.

We very much appreciate your time, and, more importantly, all you do to improve operational reliability while keeping your coworkers safe!

Sanya Mathura
Bob Latino

Authors

Sanya Mathura is the Founder of Strategic Reliability Solutions Ltd based in Trinidad and operates in the capacity of Managing Director and Senior Consultant. She works with global affiliates in the areas of Reliability and Asset Management to bring these specialty niches to her clients. Sanya possesses a strong engineering background with a BSc in Electrical and Computer Engineering and an MSc in Engineering Asset Management. She has worked in the lubrication industry for the past several years and has used her engineering background to assist various industries with lubrication-related issues locally in both Trinidad and Tobago, regionally, and internationally. She has solved lubrication problems and provided training in the Automotive, Industrial, Marine, Construction, and Transportation sectors. Sanya is the first person (and first female) in her country (Trinidad & Tobago) and by extension the Caribbean to attain an ICML MLE (International Council for Machinery Lubrication Machinery Lubrication Engineer) certification.

Robert J. (Bob) Latino is currently a Principal with Prelical Solutions, LLC (www.prelical.com). Previously Bob was a founder and CEO of Reliability Center, Inc. (www.reliability.com) until it was acquired in 2019. RCI is a 49-year-old Reliability Consulting firm specializing in improving equipment, process, and human reliability. Bob received his bachelor's degree in Business Administration and Management from Virginia Commonwealth University. Bob has been facilitating RCA analyses with his clientele around the world for over 36 years. He has taught well over 10,000 students in 25+ countries, the PROACT® Methodology, and associated software solutions. He is the author or co-author of seven books related to RCA, Reliability, Failure Mode and Effects Analysis (FMEA), and/or Human Error Reduction. Mr. Latino is an internationally recognized author, trainer, software developer, lecturer, and practitioner of best practices in the field of Reliability Engineering with a specialty in all aspects of a holistic RCA system. Bob currently serves on the Board of Directors for HPRCT (Community of Human and Organizational Learning – www.cholearning.org). He is also the Series Editor (Reliability Engineering Specialist) for CRC Press/Taylor & Francis (www.taylorandfrancis.com). His series is entitled "Reliability, Maintenance, and Safety Engineering: A Practical View of Getting Work Done Effectively".

How to Read This Text

HOW TO GET THE MOST OUT OF EFFICIENTLY READING THIS TEXT

So what is the best way for *you* to gain as much value from this book, in the most efficient manner possible? This book is not written in a sequential fashion, in the sense that you must read the first chapter in order to understand the second. There will be cross-referencing when appropriate to completely understand the principles involved.

The value to be derived depends on your existing proficiency with LDMs and RCA approaches. For novices just getting into the field, they will likely gain more from reading the entire text than veterans, who may focus on the deeper chapters that will get into the understanding of the root causes associated with specific LDMs.

Here is a summary of the contents of the book so you can pick and choose what works best for you:

1. Chapter 1 focuses on the fundamentals or basics of lubrication. This is a foundational chapter if one does not have a good handle on lubrication basics.
2. Chapter 2 focuses on the fundamentals or basics of RCA with a summary of the steps involved in the PROACT® RCA approach. Like Chapter 1, this is a foundational chapter for anyone who is not comfortable with the basic steps of any investigative occupation.
3. Chapter 3 now starts to look at the big picture of LDM and RCA and starts to integrate the principles in a synergistic manner. We start with viewing the principles from a systems and strategy perspective, and then we start to delve into tactics in upcoming chapters.

4. Chapters 4–9 go into depth for each of the six lubrication degradation mechanisms:
   a. Oxidation
   b. Thermal Degradation
   c. Microdieseling
   d. Electrostatic Discharge or ESD
   e. Additive Depletion
   f. Contamination

Each chapter is formatted similarly, where the LDM is explained in detail at the beginning of the chapter. Thereafter, there will be a logic tree for each mechanism, along with a verification log for each hypothesis in the logic tree. The intent is that each logic tree will serve as a troubleshooting flow diagram for analysts in the field to use as a reference guide. As they find hypotheses to be true, they will drill deeper. As they find them to be "not true", they will cross them off. In the end, analysts will have an RCA starter template for each of the lubrication degradation mechanisms.

Chapter 10 brings the entire book together by applying the RCA principles to a specific Fan Motor Bearing Failure. While Chapters 4– 9 create troubleshooting flow diagrams, Chapter 10 explores a singular event and applies the principles taught throughout the book.

So, we hope this quick blueprint of the book layout will help you gain the greatest value, in the least amount of time.

Thank You.

Sanya Mathura
Bob Latino

Lubrication Basics

1

CAN OIL FAIL?

Within the industry there has always been a great argument that it is not the oil that fails, it is the machine. When we think about the failure of a component, the immediate thought is that the component could not perform its specified function under normal operating conditions. If we apply this thought principle to the argument above, then we can deduce it is both the machine and the lubricant that have failed. Even though they can be classed as separate components, they are integral to each other to performing their required functions.

Ideally, any component can fail if subjected to conditions that are outside of its operating envelope. In this particular case, if a lubricant is subjected to conditions which cause it to fail, then we must investigate the source of these conditions. By following deductive reasoning, the source of these conditions can be the immediate environment or, in this situation, the machine. We must now investigate what caused the machine to be pushed outside of its operating limits. Hence, it can be said that the great argument should perhaps be rephrased to state, "oil can fail when its conditions induce such a failure".

WHAT ARE THE FUNCTIONS OF A LUBRICANT?

Let's think about a car starting at the beginning of the day. Assuming the car was not used during the night, we can expect the oil in the engine is at the bottom of the sump and everything is at rest. When the user goes to the car to start the engine, they turn the ignition (or push the button, as is the case these days

DOI: 10.1201/9781003252030-1

1

with more modern cars). The engine is immediately brought to life and all of the parts in it begin to move. It will take some time for the oil (depending on its viscosity) to get to all of the parts which require lubrication.

We can take a step back and try to mimic the moving parts in the car engine to that of our hands rubbing together. When we rub our hands together, they produce heat because of the friction of the two surfaces moving against each other. If we were to rub our hands together faster, more heat will be generated. However, if we added some cream between our hands and did the act of rubbing our hands together faster, we would realize our hands didn't produce as much heat. Our hands don't move as quickly as the pistons in an engine. Can you imagine the amount of heat being produced without any lubrication in that area?

Essentially, one of the main functions of a lubricant is to *reduce friction* between two moving surfaces. Another function is to reduce heat (as we noticed with the cream between our hands), which promoted an increase in efficiency. If our hands had some dirt on them and we still used the cream, we would notice that the dirt would be removed. If this had remained, it could have damaged our hands. Two other main functions of a lubricant include the *removal of contaminants* and *protection from wear.*

We can compare this to what occurs in an engine. The pistons will move up and down on the liner (initially without any lubricant on startup!) generating heat. When the lubricant reaches the liner, it can help in *reducing the amount of friction* produced by providing a layer upon which the piston can now easily slide. This will *reduce the amount of heat* and can *prevent wear* (since the piston is not moving over the surface of the liner by itself). Additionally, any contaminants that ingress into the system can be removed as the oil will either keep these in suspension or move them away from the areas of contact. As we noticed with the cream in our hands, the lubricant will also greatly increase the efficiency of the pistons moving across the liner.

To summarize, the *five main functions of a lubricant* are to *reduce friction, minimize wear, distribute heat, remove contaminants,* and *improve efficiency.* The functions of lubricants are not limited to these and can change depending on what is required of a lubricant. For instance, the main purpose of hydraulic oils is to transmit power, whereas if we thought about the lubricant from an analysis perspective, its purpose would be a conduit of information.

WHAT ARE THE COMPONENTS OF A LUBRICANT?

When performing Root Cause Analysis (RCA), it is imperative that we understand what we are investigating in its entirety. In the case of lubricants, this consists of understanding their functions, composition, and purpose as per application. The basic components of oil include base oil and additives, whereas the components of grease include base oil, additives, and thickener. Depending on the application of the oil or grease, components are chosen.

There are five main groups of base oils (as per API, 2015), which can be further subdivided into two main categories: mineral and synthetic (lab-based and naturally occurring). Not all the molecules in mineral oils have the same shape. However, with synthetic oils, all of the molecules are similarly shaped. Each base oil group has particular attributes which influence the characteristics of the finished product.

For instance, Group I base oils are solvent-refined and were the first type of oils to be widely used. They have more solubility than Group II base oils, which are hydrocracked or hydrotreated. In 2010, the industry underwent a change from Group I to Group II base oils. Turbine users started noticing high volumes of varnish after this base oil change as the varnish deposits were no longer able to stay in the solution as they did previously with Group I base oils. Since then, there have been significant advancements within the industry and now varnish issues are not as prevalent as they were back then. Hence, the composition of the finished lubricant is particularly important in understanding its characteristics and functions.

Table 1.1 gives a brief overview of the five main groups of base oils as per API (2015). It must be noted, however, there is a sixth base oil group as defined by the Association Technique de L'Industrie Européenne des Lubrifiants (ATIEL) in Europe. Group VI includes all polyinternalolefins (PiO).

Additives are sacrificial in nature. They are usually added to the base oil to *enhance, suppress,* or *introduce new properties*. However, different applications require diverse blends of additives. For instance, typical turbine oils are blended in the ratio of 99% base oil and 1% additives. On the other hand, most motor oils use around 70% base oil and 30% additives. This accounts for the fact that turbine oils are generally very light in color while motor oils are darker. In each case, the additives required and the blend would vary.

Through comprehension of the additives for particular oils, the user can easily identify if an additive has dangerously depleted or been dropped out

TABLE 1.1 Base Oil Groups as Defined by API (2015)

BASE OIL	SATURATES	SULFUR	VI
Group I	<90%	>0.03%	$80 \leq VI < 120$
Group II	≥90%	≤0.03%	$80 \leq VI < 120$
Group III	≥90%	≤0.03%	$VI \geq 120$
Group IV	Polyalphaolefins		
Group V	All other base stocks not included in Groups I, II, III, or IV		

to react with contaminants in the oil. In turbine oils, oxidation is a primary mode of degradation. As such, turbine oil users typically monitor the health of their antioxidants (which protect the base oil). This is usually done through the Rotating Pressure Vessel Oxidation Test (RPVOT) or the Remaining Useful Life Evaluation Routine (RULER). We will cover these tests in more detail in another chapter.

THE IMPORTANCE OF VISCOSITY

Three of the main functions of a lubricant are to reduce friction, minimize wear, and improve efficiency. The correct viscosity is critical to each of these functions. If the lubricant is too thick, it will create a viscous drag between the two surfaces, thereby reducing the efficiency. Contrarily, if the lubricant is too thin, then it will not provide adequate separation between the two surfaces. Thus, the asperities of the surfaces will touch and create friction in addition to wear.

Viscosity is deemed one of the most important characteristics of a lubricant. Achieving and maintaining the correct viscosity is critical for all applications, as the wrong viscosity can lead to the failure of equipment. Hence, it is very important to ensure the correct viscosity is used as per the application. It must also be considered that the operating conditions of equipment are not the same everywhere. Thus, these operating conditions need to be factored in when selecting the correct viscosity. This can usually be verified through the Original Equipment Manufacter (OEM) guidelines for the equipment.

One example can be using the wrong viscosity of oil in an engine. Perhaps the rating of the multigrade oil was too high (20w50) when the engine was rated for a 0w20. Taking a bit of a step back, the number in front of the "w" can be likened to the viscosity on startup, while the number after the "w" indicates the viscosity during operation. Hence the higher the number in front of the w, the longer it takes to start moving. On the other hand, the larger the number behind the "w" indicates a thicker oil. The 20w50 oil would take longer to get to the points of lubrication compared to the 0w20 on startup. Additionally, the operational viscosity of the 20w50 would cause some drag compared to the 0w20 for which it was designed.

We can think of a straw in a glass pulling up some liquid. The 20w50 can be likened to a very thick liquid such as molasses while the 0w20 is likened to water. If the size of the straw remained the same, then it would take more effort to pull the molasses (20w50) through the straw compared to pulling water (0w20). Therefore, if the car continues to use the 20w50 oil it would strain the pump, cause more gas (or diesel) to be used to power the system, and eventually lead to an engine failure.

One point to note would be that the size of engines has drastically reduced over the years. If the size has been reduced, then the oil line clearances would have been reduced as well. Therefore, using a 20w50 oil (with a larger oil line clearance) would have been perfectly fine more than 30 years ago. Since clearances have been reduced, then the viscosities required by newer vehicles have to be reduced as well. We can think back to the straw example, if we had a thinner straw then it would be more difficult to pull through the molasses (20w50) compared to water (0w20).

UNDERSTANDING THE LUBRICATION REGIMES

Each piece of equipment undergoes at least three types of lubrication regimes in its lifetime. These regimes occur on startup, shutdown, and during operation. It is important to understand the regimes so that we can identify if this can be a possible contribution to a failure mechanism. For instance, during startup, "Are elevated temperatures seen which continue throughout the operation of the equipment, or do these temperatures decline after some time?"

The four regimes are boundary lubrication, mixed lubrication, hydrodynamic lubrication, and elastohydrodynamic lubrication as per Wang and Wang (2013, 184).

Boundary lubrication usually occurs during startup or shutdown of the equipment. In these instances, the lubricant does not form a full film layer to adequately separate the asperities from both surfaces. Therefore, considerable wear occurs during this phase.

Mixed lubrication occurs after the equipment has started and the film begins to form. However, it is still in the process of being formed and not all areas of the film are of the same consistency. During this phase, there are some parts of the surfaces which experience a full film but there are also other parts that experience boundary lubrication.

Hydrodynamic lubrication occurs during the operation of the machine. In this regime, both surfaces are adequately separated and there is no touching of asperities. The lubricant makes a full film and helps in reducing friction and improving the efficiency within the component.

Elastohydrodynamic lubrication is truly a special regime as it elastically deforms the surface of the component at the highest point of contact pressure. This is done so that the asperities of the two surfaces do not touch while maintaining a full film of lubricant. Typically, this occurs during the operation of the component and will not be seen during startup or shut down of the equipment.

MECHANISMS OF WEAR

We are now more aware of the various lubrication regimes and the importance of viscosity. It must be noted in these cases that wear occurs between the two surfaces and varies depending on the conditions present. As such, a sound understanding of the different types of wear mechanisms must be acquired before we launch into the breakdown of larger components.

As per Bhushan (2013, 315–316), wear occurs when there is surface damage or removal of material from one or both of two solid surfaces which are in sliding, rolling, or impact motion relative to each other. Some may argue wear only takes place when there is a net volume loss to either surface. However, we must understand that during initial contact between the two surfaces, the material on one surface may become altered without loss of material but during further contact, the material is lost. Additionally, if two surfaces transfer material to each other, there can be a net-zero loss of material although wear has occurred.

There are six main distinctive mechanisms of wear namely *adhesive, abrasive, fatigue, impact by erosion* and *percussion, chemical* (or *corrosive*), and *electrical-arc-induced wear* as per Bhushan (2013, 315–316). Each wear mechanism produces different characteristics, which can provide the user with more information on the process by which the wear occurred. While other forms of wear can exist, they are usually combinations of these six forms. Interestingly enough, two-thirds of industrial wear are attributed to adhesive or abrasive wear mechanisms.

Adhesion wear occurs when two bodies slide over each other and the asperities become bonded through a weld. As the sliding of the bodies continues, the weld is broken, and this can result in fragments being broken off or material being transferred from one surface to another. Sometimes, the process is repeated in a cyclical pattern that allows for the buildup of the transferred material.

Abrasive wear is slightly different. It involves the plastic deformation or fracture of the softer surface. In this type of wear, there is usually a harder, rougher surface and a softer surface. The harder surface causes the softer surface to become plastically deformed until it reaches the point of being fractured. Normally, scratches, gouges, or scoring are seen on the surface. There are two types of abrasion: two-body and three-body abrasion.

Two-body abrasion normally occurs in mechanical operations such as grinding, cutting, and machining. *Three-body abrasion*, on the other hand, involves a third particle which meets two other surfaces. This third body is usually harder than the other two and can remove material from either one or both surfaces. An example of this type of abrasion can be found in polishing. However, it must be noted that for this third-body abrasion, adhesion may be the first form of wear.

Fatigue wear is very different from adhesion or abrasion. Fatigue wear experiences repeated rolling and sliding friction on both the subsurface and surface layers. After repeated cycles, where there is no loss of material, a crack is formed and the surface is broken sometimes forming spalls. This results in large fragments being removed and gives rise to the formation of pits, commonly called pitting. Adhesion and abrasion undergo gradual deterioration while fatigue experiences cyclical stresses without loss of material until the surface becomes broken.

Impact wear can be classified as either erosion or percussive wear. *Erosion* involves the removal of material from a surface through jets and streams of particles, liquid droplets, or implosion of bubbles. On the other hand, *percussive wear* happens due to repeated impacts of solid bodies to remove material from the surface.

Two conditions need to be present for *chemical or corrosive wear*. Sliding between the two surfaces must occur within a corrosive environment. The sliding action removes any protective coating, revealing the surface to the corrosive environment. Hence, corrosion is normally very dominant in areas exposed to highly corrosive environments such as gases in industrial plants.

If there is a high positive charge over a thin air film during the sliding process, this can form a dielectric and can cause arcing also known as *electrical-arc-induced wear*. This can lead to large craters being formed. As such, any sliding or oscillating motion can fracture the lips of the crater. Consequently, this can lead to three-body abrasion, corrosion, surface fatigue, and fretting.

Fretting, although it is not one of the main six mechanisms, is a combination of adhesive and abrasive wear. This occurs when there is a low-amplitude oscillatory motion in a tangential direction between two surfaces at rest. Typically, this can occur due to vibration of equipment or during the transportation of the equipment. The normal load causes adhesion between the asperities of the two surfaces. Then, the oscillatory movement causes ruptures which result in wear debris, giving rise to three-body abrasion.

Since we understand the six main wear mechanisms, we can use these as additional artillery when performing our RCAs. These mechanisms are easily identifiable and can give the observer more insight into the internal movements of the contact surfaces. From this insight, the observer can easily establish any correlations to the lubrication degradation mechanisms and the conditions which exist.

PROACT®
Root Cause
Analysis Basics

2

WHAT IS ROOT CAUSE ANALYSIS?

The answer depends on who we are, where we work, and which regulatory requirements we are held accountable for. Not trying to be evasive, but just realistic. Think about your own experience, and if you polled different people from different industries as to "What is Root Cause Analysis (RCA)?" What do you think you would hear? Would you envision complete agreement? Maybe we could settle on just getting a consensus.

Trying to make everyone happy is a futile effort, and to be honest, it is not necessary. What is necessary is to produce verifiable, bottom-line results and to let our performance do the talking. Do not get mired in the complexities and debates about menial disagreements on labels, brands, and legacy approaches – *focus on results.*

When we think about RCA, let's just look at it from a macro view and the rest will fall into place as you read this opening chapter. When we look at any investigative occupation, these five steps are critical to the success of the investigation (DOE-NE-STD-1004-92).*

 i. Data Collection
 ii. Assessment
 iii. Corrective Actions
 iv. Inform
 v. Follow-up

---

* U.S. Department of Energy. Office of Nuclear Energy. Office of Nuclear Safety Policy and Standards. *Root Cause Analysis Guidance Document*, 1992 http://www.hss.energy.gov/nucl earsafety/ns/techstds/standard/nst1004/nst1004.pdf

As we progress through this text, we will align with the steps of our preferred PROACT® RCA methodology, which is consistent with the steps of any investigative process.

For the purposes of this text, we will use the PROACT® definition of Root Cause Analysis,* which is:

> The establishing of logically complete, evidence-based, tightly coupled chains of factors from the least acceptable consequences to the deepest significant underlying causes.

This is a variation of a definition that was proposed on the Root Cause Analysis discussion forum at www.rootcauselive.com decades ago.†

While a seemingly complex definition, let us break down the sentence into its logical components and briefly explain each:

1. Logically Complete – This means all the options (hypotheses) are considered and either proven or disproven using hard evidence.
2. Evidence-Based – This means hard evidence is used to support hypotheses as opposed to using hearsay and treating it as fact.
3. Tightly Coupled Chains of Factors – These fancy words mean we are using cause-and-effect RCA approaches as opposed to categorical (cause categories where we brainstorm disconnected possibilities in each category) RCA approaches.
4. Least Acceptable Consequences – This is the point where the event that has occurred is no longer acceptable (a trigger of some sort has been hit) and an investigation is launched.
5. Deepest Significant Underlying Causes – These seemingly intimidating words mean at what point do we stop drilling down and determine going deeper adds no value to the organization?

This definition certainly encompasses and embodies the intent of the Department of Energy (DOE) guidelines for RCA as shown in Figure 2.1.

---

* PROACT® is a registered trademark of Reliability Center, Inc. Used with permission.
† This discussion forum is associated with www.rootcauselive.com and moderated by Mr. C. Robert Nelms.

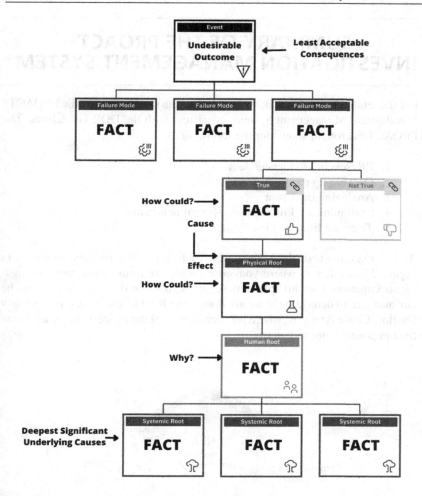

FIGURE 2.1   PROACT® RCA summary logic tree.

SUMMARY OF THE PROACT® INVESTIGATION MANAGEMENT SYSTEM*

For the remainder of this text, we will follow the simple steps of the PROACT® Investigative Management System, consistent with the DOE Guidelines. The PROACT® acronym stands for the following:

1. **PR**eserving Evidence/Data
2. **O**rganizing the RCA Team
3. **A**nalyzing the Event
4. **C**ommunicate Findings and Recommendations
5. **T**racking Bottom-Line Results

As you read this text, think about how each step of the process, as shown in Figure 2.2, applies to where you work. When we think about any investigative occupation, they all follow these steps to some degree. This should be our measure to determine if we are doing true Root Cause Analysis or simply Shallow Cause Analysis, where we succumb to taking process shortcuts due to time pressures imposed on us.

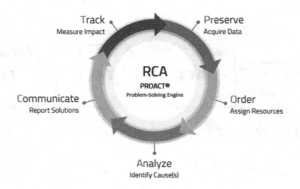

FIGURE 2.2   PROACT® RCA methodology cycle.

---

* Latino, Robert, Kenneth C. Latino, and Mark A. Latino. 2019. *Root Cause Analysis: Improving Performance for Bottom-Line Results,* 5th ed. Boca Raton, FL: CRC Press.

WHY DO UNDESIRABLE OUTCOMES OCCUR? THE BIG PICTURE

We must put aside the industry that we work in and follow along from the standpoint of human beings. In order to understand why undesirable outcomes exist, we must understand the mechanics of failure.

Virtually all undesirable outcomes are the result of human errors of omission or commission (or decision errors or choices, as they will be used interchangeably from now on). Experience in industry indicates that any undesirable outcome will have, on average, a series of 10–14 cause-and-effect relationships that queue up in particular patterns for that event to occur.

This dispels the commonly held myth that one error causes the ultimate undesirable outcome. All such undesirable outcomes will have their roots embedded in the physical, human, and latent areas.

Physical Roots: These are typically found soon after errors of commission or omission. They are the first physical consequences resulting from the choices made. Physical roots, as will be described in detail in later chapters, are tangible things we can see.

Human Roots: These are decisions that are made that did not go as planned. These are the actions (or inactions) that trigger the physical roots to surface.

Latent Roots: These are the organizational systems that are flawed in some manner (i.e., inadequate, insufficient, and/or non-existent). These are the organizational support systems (i.e., procedures, training, incentive systems, purchasing habits, etc.) that are typically put in place to help our workforce make better decisions. Latent roots are the expressed intent of the human decision-making process.

WHAT ARE THE ELEMENTS OF A TRUE ROOT CAUSE ANALYSIS "SYSTEM"?

To recognize what is Root Cause Analysis and what is *not* Root Cause Analysis (Shallow Cause Analysis), we would have to define the criteria that must be met in order for a process and its tools to be called Root Cause Analysis. In the absence of a universally accepted standard, let us consider the following essential elements* of a true Root Cause Analysis process:

1. **Identification of the *Real* Problem to Be Analyzed in the First Place.** About 80% of the time, we are asked to assist on an investigation team – the problem presented to us is not the problem at hand.

2. **Identification of the Cause-and-Effect Relationships that Combined to Cause the Undesirable Outcome.** Being able to correlate deficient systems directly to undesirable outcomes is critical.

3. **Disciplined Data Collection and Preservation of Evidence to Support Cause-and-Effect Relationships.** It is safe to say that if we are not collecting data to validate our hypotheses, we are not properly conducting a comprehensive RCA.

4. **Identification of All Physical, Human, and Latent Root Causes Associated with Undesirable Outcome.** If we are not identifying organizational system deficiencies and contributing cultural norms that lead to poor decision-making, then, again, we are not properly conducting a comprehensive RCA.

5. **Development of Corrective Actions/Countermeasures to Prevent Same and Similar Problems in the Future.** If we have merely developed good recommendations but never implement them, then we will not be successful in our RCA efforts. This is where the ball is often dropped as well-intentioned people are pulled away by reactive work, and these proactive opportunities fall by the wayside.

6. **Effective Communication to Others in the Organization of Lessons Learned from Analysis Conclusions.** One of the greatest benefits of a successful RCA is the dissemination of the lessons learned to avoid recurrence elsewhere in the organization. Oftentimes, successful analyses end up in a paper filing system only to be suppressed from those who could benefit from the lessons learned in the analysis.

This mini-book is intended to streamline the understanding of the PROACT® RCA process for those that must do the work. So, we will avoid as much theory as possible and outline the basics associated with "how to" conduct more, better, and faster (more efficient) RCAs.

The core of the PROACT® RCA methodology is undoubtedly the creation of what we will call "the logic tree". This is our event reconstruction tool of

---

* Latino, Robert J. January 2005. "PROACT Approach to Healthcare Workshop. " www.proact-forhealthcare.com

choice. Again, look at our steps in the reconstruction process for their objectives, as opposed to feeling like you must do it the same way. In the end, we just want to get similar results even though we got there through different means.

FUNDAMENTAL ELEMENTS OF A LOGIC TREE

Cause-and-Effect Logic: From level to level in a logic tree, it represents a cause-and-effect relationship. This does not have to be a linear relationship as there may be multiple causes that have to occur at the same time to create that effect. We just need to know that we are simply creating a graphical expression of logic to reflect the facts that occurred in order to cause an undesirable outcome.

Event: Simply put, "The reason you care"! What brought this incident to your attention? Many believe that we do RCA on incidents themselves, but we believe we do RCA on their consequences. Think about it at your place; there is usually a business level reason we do RCA – injury/fatality, a certain amount of production loss exceeded, a certain amount of maintenance cost exceeded, regulatory violation, and the like (often called triggers). Does it sound familiar? These are the known *facts*.

Failure Mode(s): These are the typical things we normally start an RCA with, like pump failure, injury, loss of production, Environmental Excursion. These led to the Event. These too are *facts*. The term MODE can be confusing because of its usage in approaches like RCM and FMEA. However, an easy way to remember the difference is that a Reliability Centered Maintenance (RCM)/ Failure Mode Effect Analysis (FMEA) mode is something that "could" occur (a risk of occurring). An RCA Mode means that RCM Mode *did* occur.

Hypotheses: Just like in high school, these are "educated guesses". These are potential causes to the preceding nodes. The initial question after the Failure Mode is "How could the preceding node have occurred?"

Verifications: These are the ways in which we proved, with sound evidence, that hypotheses were true or not true. Fun fact – hearsay is *not* a valid verification technique.

Physical Root Causes: These are where the physics of failure root out. These are observable, tangible things we can see. These are usually the immediate consequences of decision errors.

Human Root Causes: *This is not "the who dunnit"*! This is the act of decision-making. These are usually errors of omission and/or commission. We did something we were not supposed to, or we were supposed to do something and did not. The key here is to *not blame*, and we should take the opportunity to understand human reasoning (*why* people do what they do).

Latent/Systemic Root Causes: These are the organizational systems, cultural norms, and sociotechnical factors that influence and contribute to our decision-making. Unfortunately, our "systems" are far from perfect and are always a work-in-progress. They include, but are not limited to, our policies, procedures, training systems, purchasing systems, HR systems, compliance systems, enforcement systems, and the like.

Contributing Factors: These identify items that did not directly lead to the failure but created vulnerabilities allowing the failure to occur. These are usually conditions that we do not have control over, but we can often compensate for them (if we are aware of them). For instance, some failures may only occur when it is freezing outside. This is a condition that we cannot change, but we can compensate for them in order to mitigate their potential consequences.

OK, now let's put these pieces of the puzzle together and show how to reconstruct an undesirable outcome!

In Figure 2.1, as mentioned earlier, the *event* is the reason we care enough to commission an RCA. In our example, the event is "Unexpected Failure of the Turbine Generator". Now the *mode* is going to be how we have experienced this failure in the recent past (there can be multiple *modes*). Most of our CMMS's can produce these types of high-level modes. In this case, such downtime due to this turbine failure has been attributed to the failure of critical bearings. Our data also can tell us the annual cost of each of these modes (such as downtime cost + labor cost + materials cost). To make a quick business case for your failure, try out this free Chronic Failure Calculator (CFC) at https://www.reliability.com/.

The mode level is what we consider our *fact line* to start as shown in Figure 2.3. If we start with facts and provide our hypotheses with sound validations, we will end with facts. Keep in mind we are traveling down the path of the physics of failure, and so we will continually ask the same question, "How Could".

As you use a logic tree to explore the physics of failure, imagine you have the luxury of a video recorder in your head and you are watching the event as it is played in reverse. In our case, "be the bearing". Ask yourself, "How could I have just failed?" Move back in short increments of time. It takes some getting used to this type of thinking, but that is the beauty of the logic tree, it guides us without any biases. This tool, when used properly, should be nonpersonal and nonthreatening. We are interested in valid hypotheses, possibilities – that's it. Then, we will use evidence to demonstrate which hypotheses were true and which were not true. We will only continue drilling down on the ones that are true.

Based on the subject matter Expert (SME) on our team, we conclude that based on the evidence in the case, our hypothesis is "Progress of Impending

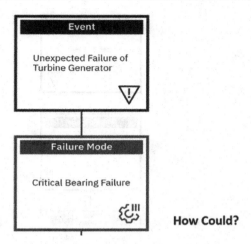

How Could?

FIGURE 2.3   Event + mode(s) = top box (all must be *facts*).

Bearing Failure Went Undetected (Infrequent Oil Analysis)". So, we list them as shown in Figure 2.4.

In our example, we review the oil analysis frequency records in an attempt to validate this hypothesis. We determine with certainty from this review that the bearing was slowly starved of by reduced and restricted oil flow. This just begets another question, "How could the bearing have slowly starved of by reduced and restricted oil flow?"

Our SME indicated either from "Spongy Entrained Air Reducing the Pumping Flow" and/or from "Varnish Deposits Formed in the Oil Ways". The visual inspection confirms both to be true.

The path forward is pretty simple; our team asks, "How can we have 'Spongy Entrained Air Reducing the Pumping Flow'?" and "How can we have 'Varnish Deposits Formed in the Oil Ways'?" Both can be validated by visual inspections. For the spongy entrained air, we can verify this by inspecting the sump for the presence of foam or opening the pump and inspecting it for cavitation inside. For the varnish deposits, verification can take the form of visual inspection, increases in temperature (due to the insulating effect of varnish), or vibration analysis particularly in areas with a shaft. The vibration analysis will show the misalignment of the shaft.

Now that we've established the questioning and validation protocols of the cause-and-effect logic, it becomes pretty easy to follow the logic graphically using the logic tree to tell our story.

As we can see, questions only beget more questions, as that's what effective RCA analysts do for a living; they ask the right questions!

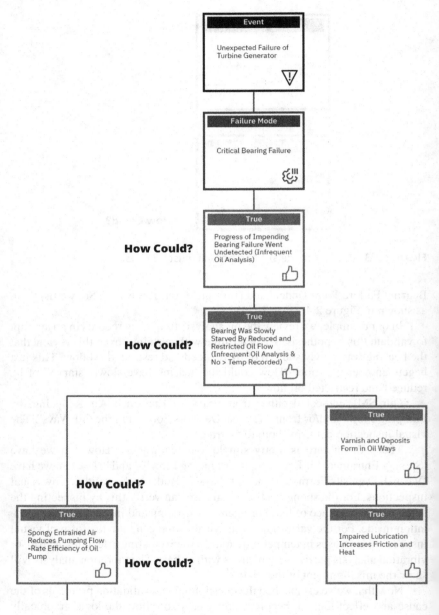

FIGURE 2.4   Hypothesizing and validation.

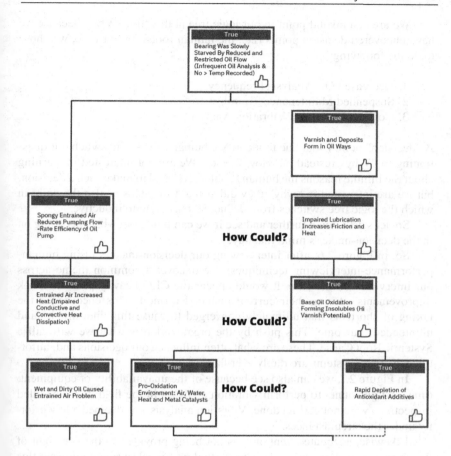

FIGURE 2.5   Continued hypothesizing.

In Figure 2.5, this is where we are now crossing over from the physics of failure to the human and systems' side of failure (or the social sciences). Notice here that we switched the label on the "Aerated, Wet and Metal Particle-Laden Oil Went Undetected for Several Weeks" node from a hypothesis to a physical root. This is because this is the first visible consequence after the triggering decision.

Notice that after the decision point, everything is triggered on its own, as cause-and-effect linkages go into play. If there are no human interventions to break the error chain, then it will play out and contribute to the undesirable outcome (event).

We are at a pivotal point in our logic tree at this time. Why? Because we have uncovered decision points (i.e., the human roots). In our case, we chose to do the following:

1. Decreased Oil Analysis Frequency
2. Suspended Maintenance Training
3. Conduct Infrequent Vibration Analysis

A "decision" point is our cue to identify a human root and to switch our questioning to "Why" instead of "How Could". We are not interested in learning about the infinite reasons the human "could have" had to make such a decision, but we are interested in "why" they did at that time. This is also the point in which the logic tree switches from deductive reasoning to inductive.

So, let's drill down further and see if we can figure out what was going on in the decision-maker's mind that day!

So, in Figure 2.6, after interviewing our decision-makers (using human-performance-interviewing techniques), we uncover a common theme across our interviews in this case. It would appear the CEO heavily pushed "quick improvements" during their current financial turmoil. This encouraged the taking of shortcuts that ultimately converged to cause this undesirable (and unintended) outcome. This push by the organization is what we will call a Systemic Root Cause. These are what often influence our decisions and, unfortunately, these systems are rarely ever flawless.

In Figure 2.7, we can also see because of the unavailability of equipment, time, and personnel to perform vibration analyses in the field, the scheduled inspections were not getting done. Vibration analysis was deemed a low priority under the circumstances.

Likewise, adequate attention was not being provided to the oversight of the alarm systems, which could have alerted personnel in time to prevent this failure from materializing.

Let's follow the same logic as described in Figure 2.6 and apply it to Figure 2.7 to round out this case study. Let's assume we validate with hard evidence that we definitely had "Growing Dirt Concentration in the Oil". When exploring "how that could happen", we find out the following decisions were made, which could be the causes:

1. We Bought Cheap Filters (a choice)
2. Bought Cheap Turbine Oil (Base Oil and Additives) Which Could Not Protect the System (a choice)

We stick to our RCA discipline and ask "Why" were these decisions made that day? Everything leads to the CEO mandate that encouraged taking risks and

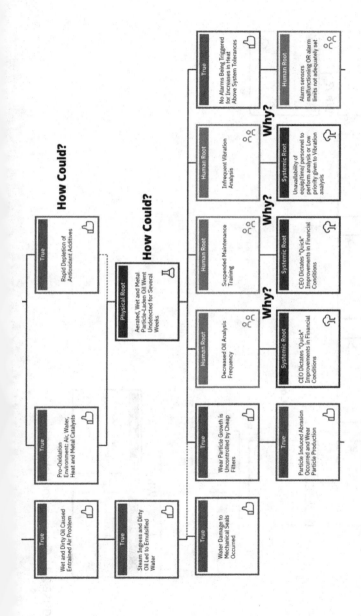

FIGURE 2.6 Continued hypothesizing and root labeling.

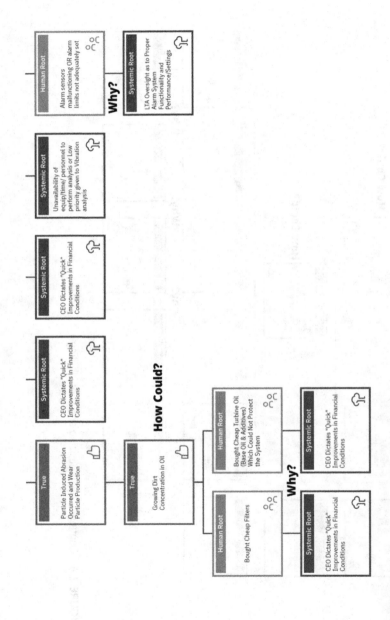

FIGURE 2.7 Continued root labeling.

shortcuts, and in this case, it didn't pan out and ended up having the opposite effect. We felt we were doing what the CEO wanted us to do, and so our decisions were well-intended.

We find out that the trade-off pressures between Finance/Operations and Maintenance/Reliability influenced the decision. There was significant pressure to keep the process running again while minimizing the costs. When that typically happens, the natural human tendency is to take shortcuts. This usually comes in the form of skipping steps in a sequence of tasks we must do.

Chances are that these "Systemic Roots" have contributed to other failures as well individually or in combinations. This is because most "systems" are put in place for a multitude of people to use, under a variety of conditions. This mandate resulted in the convergence of consequences on this day, to influence the well-intended decisions that day.

When Do We Stop Digging?

Our personal rule of thumb to answer this question is, "When the solution is obvious"! If drilling deeper gets into issues outside our fences, is it of value? In most cases, we will not have control over things like changing regulations. However, in many cases, if we see an Original Equipment Manufacturer (OEM) design flaw, we may opt to have our engineering department pursue that path with the OEM. But as the analyst in the field, we can hand that part off and move on to my next RCA.

Is everything about every facet of "RCA" covered in this chapter? Absolutely not. This is why our title includes the word *Basic*.

Remember, "We NEVER seem to have the time and budget to do things right, but we ALWAYS seem to have the time and budget to do things again!" Let's do RCA right the first time so we don't have to analyze the same event again 😊!

The Application of Root Cause Analysis (RCA) to Lubrication Degradation- Related Failures

3

THE *BIG* PICTURE: VIEWING "RCA" AS A SYSTEM

This chapter is not going to debate what "RCA" is and contrast the differences in the many different approaches in the marketplace. However, we can get away from RCA labels and brands and generically break down the critical steps of any investigative occupation. There is no need to get mired down in the weeds of "distinctions without a difference", as this is a Focus Book (how-to) and not an academic or research book.

In our experience, we would estimate that about 75% of those who practice whatever they consider RCA view RCA as a task and not a system. So, when an unexpected failure occurs that interrupts operations (or other business conditions), and someone in authority will demand to know what happened and commission an RCA. As we all know though, in practicality, the first question those authorities will ask is "When will we be back up and running?" and "How much will it cost?". Those are the harsh realities of our field of choice,

DOI: 10.1201/9781003252030-3

and that is exactly what drives "RCA" to be viewed as a task. Under such conditions, conducting an RCA is totally reactive in nature and a "one-off" task. The process is as follows: (1) failure occurs, (2) conduct RCA, (3) implement corrective actions, and (4) move on.

We are seeking to broaden the understanding of what an effective RCA "system" looks like. We are more interested in how to leverage the successful logic of all our RCAs across the company and incorporate them into a single RCA knowledge management system. We are seeking to truly be a learning organization, as opposed to being a crisis management center.

Think of how much time and resources we waste doing RCA rework. This occurs when we have similar failures at different sites, within the same company, but no one knows a successful RCA was already conducted. Therefore, we do another RCA. When we pose this question to our classes and audiences, "How much do you think an average RCA costs when a team is assembled after a serious incident (taking into consideration employee time, consultant time, software used, conference room times, interviews, collecting evidence, and the like)?" – the average we hear is 25,000 USD. Multiply that number by the average number of RCAs that are repeated on the same type of incident, and you will get your CFO's attention. CFOs are very interested when a solid business case is presented to them. Here is a resource for doing a 30-second calculation for making a sound business case using this Chronic Failure Calculator.*

To be effective investigators/analysts, we must have a grasp of the bigger picture of RCA when it comes to understanding how and why failures occur. To that end, we will briefly discuss the below:

I. Every Process Is Part of a Bigger System
II. All Outcomes Are a Result of a Series of Cause-and-Effect Relationships
III. We Can See Bad Outcomes, but We Can't See Human Reasoning (Thought Processes)

---

* Accessed 3.21.21 at https://www.reliability.com/. Reliability Center, Inc. Chronic Failure Calculator.

RCA AND SYSTEMS THINKING

Successful RCA analysts are "Systems Thinkers"! In a paper entitled "Get to the Root of Accidents",* the noted authors cite that systems thinking is not currently used in RCA.

We never like to make such definitive statements, but, in general, we would have to agree that most of what we see around the world in terms of what people call "RCA", there is little appreciation for the system's thinking.[†]

This article defines systems thinking as:

Systems thinking is an approach to problem solving that suggests the behavior of a system's components can only be understood by examining the context in which that behavior occurs. Viewing operator behavior in isolation from the surrounding system prevents full understanding of why an accident occurred — and thus the opportunity to learn from it.

It's hard for us to see how an RCA can be effective without such systems thinking. If an RCA analyst is not exploring the surrounding environment of a decision-maker for proper context to their decision-making, we would not consider that an RCA. We would consider that a "Shallow Cause Analysis"[‡] because we think they stopped short of the true systemic root causes that influenced the decision-makers.

For simplicity's sake, and not getting into a tone of theory from research/ academia, just visualize any process as a system. A "system" is composed of *Inputs* of some kind; they are *Transformed* by the process in some way, and then they produce an *Output* in the end. Think about wherever you work and put that mental image in your mind.

From the holistic RCA perspective, the system looks like this:

1. **Inputs** = Evidence collection and preservation
2. **Transformation** = Undesirable outcome analysis, reconstruction, and identification of systemic root causes

---

* Levenson and Dekker (2014). Accessed 3.21.21 at https://www.chemicalprocessing.com/artic les/2014/get-to-the-root-of-accidents

† Latino (2019). Accessed on 3.21.21 at https://www.linkedin.com/pulse/systems-thinking-cr itical-root-cause-analysiss-rca-success-latino/

‡ Latino, 2019. Accessed on 3.21.21 – https://www.linkedin.com/pulse/root-cause-vs-shallow-an alysis-whats-difference-robert-bob-latino/

3. **Outputs** = Development, assignment, and implementation of effective corrective actions and tracking of bottom-line performance improvements

ALL OUTCOMES ARE A RESULT OF A SERIES OF CAUSE-AND-EFFECT RELATIONSHIPS?

There are few things in life that you can be absolutely sure of, but one of them is that nothing ever just happens. There is always a cause-and-effect relationship. – R. Keith Mobley*

Undesirable outcomes don't just happen! This statement is gospel to a veteran RCA analyst who is relentless in the pursuit of a pattern or sequence of cause-and-effect relationships that queued up on any given day to produce an adverse outcome.

Figure 3.1 shows an example of this cause-and-effect logic when applied to a lubrication-degradation-failure scenario. Level-to-level we just ask ourselves "'How Could' the previous block have occurred?" Our answers are simply our hypotheses (possibilities) that must be validated as true or false using sound evidence (not just hearsay).

In this case, our question is "How could we have had a critical bearing fail due to lubrication degradation?"

Our potential hypotheses are:

1. Microdieseling
2. Contamination
3. Oxidation
4. Electrostatic Spark Discharge
5. Thermal Degradation
6. Additive Depletion

This was a quick demonstration of how to read logic when analyzing a failure using a logic tree for Event reconstruction.

---

* Mobley (2020). Accessed on 3.23.21 at https://www.linkedin.com/pulse/random-failures-myth-reality-r-keith-mobley/?trackingId=zH4agqgybTLHnYMQhWLKHg%3D%3D

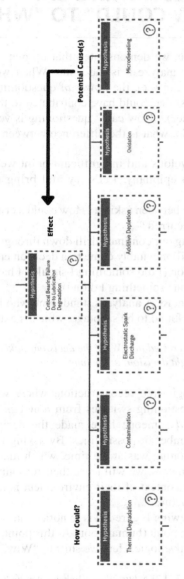

FIGURE 3.1 Demonstration of cause-and-effect relationships.

THE QUESTIONING SHIFT FROM "HOW COULD" TO "WHY"

In our previous example, we demonstrated that no matter how bad the outcomes, the structure to analyze it is the same. When we look at the physics of failure (what we can see), the *how can* questioning will reveal many possibilities (hypotheses) that could have contributed to the last effect (in the cause-and-effect sequence). "How can " questioning is very broad compared to "why" questioning. So, what is the difference between *how can and why** questioning?

It seems like a frivolous and insignificant point we're trying to make, but it's really quite an epiphany. Let's try and bring this home with an example.

Is there a difference between asking, "How could a crime have occurred?" versus "Why a crime occurred?"

At some point during our continual drill down through the physical aspect of a bad outcome, we will inevitably come to a decision error. This will either be an error of commission (I did something I shouldn't have) or be an error of omission (I should've done something I didn't).

From an independent RCA analyst's standpoint, who made the decision is irrelevant (unless it was found to be sabotage, which is extremely rare).

> *What we are more interested in is WHY the decision-maker felt the decision they made was the right decision, at the time!*

At this point in our logic tree reconstruction, where we stumble across a decision-maker, our questioning switches from *how can* to *why*. We are not interested in *how could* someone have made the decision, because there would be an infinite number of possibilities. By asking *why*, we are seeking to learn what their reasoning was at the time, which made it seem to be the right decision. Almost always, we will learn their reasoning at the time made perfect sense, when the context of their environment is taken into account – understanding the big picture.

The key point in viewing Figure 3.2 is to notice that we ask "How Could" until we reach a decision point (human root). At this point, we shift from using inductive logic to deductive logic. Here we shift to "Why" questioning. In this

---

* Latino (2021). Accessed on 3.21.21 at https://www.linkedin.com/pulse/how-can-v-why-whats-difference-robert-bob-latino/

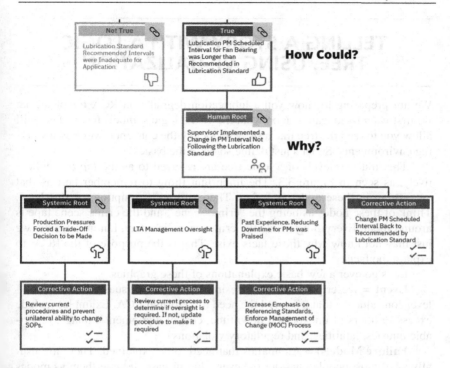

FIGURE 3.2 Shifting questioning from "How Could" to "Why".

case, we would ask "Why would the supervisor have implemented a change in Preventive Maintenance (PM) interval, thus not following the lubrication standard at the time?"

Related to this example, we find the following from our interviews as the reasoning for such a decision (remember, we cannot see "reasoning" occurring):

1. Production pressure forced a trade-off decision to be made.
2. Less than adequate (LTA) management oversight.
3. Past experience. Reducing downtime for PMs was praised in the past.

TELLING A STORY WITH A LOGIC TREE, USING VISUALIZATION

We are preparing to show you a lubrication degradation RCA template, but we first want to educate you on what the visual logic symbols mean. This will allow you to read the tree like a story and apply the concept to your own working environments. So let's just define some of the basics.

The first two levels of a logic tree are referred to as the *top box* collectively, as seen in Figure 3.3. The important thing to remember here is that everything in these nodes is *a fact*. There are no assumptions at this level. Think of these nodes as being the "crime scene", and the crime scene tape is around them. Everything inside the crime scene is a *fact*, but at the time, we just may not know why those facts exist. That is the purpose of the RCA, to explain the facts.

Let's go over a few basic explanations of these graphics.

Event = Description of an undesirable outcome. Usually is a business-level consequence that triggered the need for a formal RCA. Examples include excessive hours of unexpected downtime, cost of equipment damage, reportable injuries, fatalities, and regulatory violations.

Failure Mode(s) = Anomalies that need to be explained. These are usually what most people consider the event. In our case, we cite them as modes and they are typically the equipment that has failed (in lubrication-related cases), leading to the business-level loss.

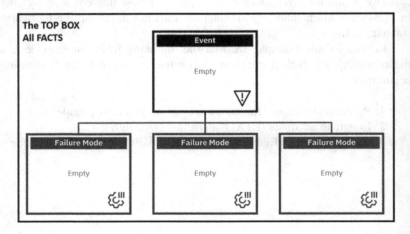

FIGURE 3.3　A logic tree top box.

For future reference, just take note now, that the perimeter of these nodes are solid lines, visually indicating they are facts, as shown in Figure 3.4.

Hypotheses = Educated guesses (just as we learned in high school). Remember, because we are exploring the physics of failure at this point, we are asking "How Could". This means we are most likely to come up with multiple hypotheses as to how something could happen. Also, remember that it is also very likely that more than one hypothesis will be true. Because failure is not linear, it is very likely that parallel paths of failure converged at some point in time to produce an adverse outcome.

In Figure 3.4, we are demonstrating the critical importance of evidence to any investigation and analysis. In Figure 3.4, see that the borders of the hypotheses were dotted lines. Now we collected evidence to prove or disprove our hypotheses and these are facts. When evidence is applied, we will visually designate the hypotheses as being *true* or *not true* using solid perimeter lines, and the application of a paper clip icon, indicating that evidence is linked to that record and ready for review. Figure 3.4 also indicates the logic tree root designations.

Physical Roots = These are typical where traditional Failure Analysis (FA) ends and Root Cause Analysis (RCA) picks up. These are using physical failure mechanisms related to failed components. From a logical standpoint, they are the first observable consequences of decision errors.

Human Roots = These are the acts of decision-making. It is the reasoning that goes on in the human mind and someone decided to do, or not do, something. In holistic RCA, unless it is sabotage (malice with intent), it is irrelevant who made an inappropriate decision. A progressive organization will not seek to blame the decision-maker and cite a cause of "Human Error". They will seek to find out why that well-intentioned person thought it was the right thing to do.

Latent or Systemic Roots = These are the reasons why the decisions were made at the human root level. These are the basis for what people believed to be true at the time. These are usually flawed organizational systems (e.g., policies, procedures, training systems, and purchasing systems), cultural norms (practices that devolved from set standards), and sociotechnical factors (issues outside the fence like regulations, insurance, and warranty issues).

Corrective Actions = These are the remedies to mitigate or eliminate the identified root causes. There is not a 1:1 ratio when it comes to corrective actions per identified root cause. There could be several corrective actions for one root cause or there could be one corrective action that will address several root causes.

Note: All logic tree graphics used with permission from Reliability Center, Inc. For more information on their EasyRCA solution, please visit https://www. easyrca.com/.

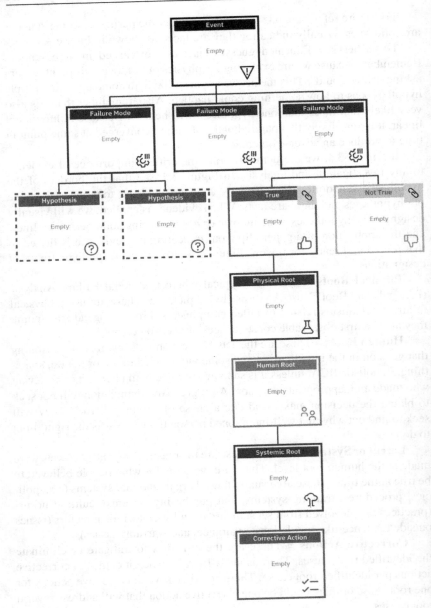

FIGURE 3.4 Developing and designating hypotheses and roots.

Building a Tree for the Lubrication Degradation Mechanism: Oxidation

4

WHAT IS OXIDATION?

Oxidation is one of the most popular forms of degradation. It attracts a lot of attention and has been used to generalize all forms of degradation. In fact, oxidation occurs only when there are significant temperature increases, with the lubricant being in the presence of oxygen. As per Ameye, Wooton and Livingstone (2015), the definition of oxidation has been expanded to include any reaction in which electrons are transferred from one molecule.

During oxidation, the initiating factor is the presence of free radicals. Free radicals are molecular fragments having one or more unpaired electrons which are accessible and can easily react with other hydrocarbons as per Ameye, Wooton and Livingstone (2015). When we think about the molecular structure of a lubricant, it is mainly composed of hydrocarbons. Hence, the presence of free radicals would readily react with these hydrocarbons.

In the *Initiation phase*, the free radical is formed. Once formed, it then produces more free radicals and moves into the second phase of *Propagation*. During this phase, antioxidants begin to react with the free radicals to make them more stable. This leads to the last stage called *Termination*. Depending

DOI: 10.1201/9781003252030-4

on the amount of free radicals, the antioxidants could be drastically depleted. Once the antioxidants have been depleted, this leaves the base oil unprotected. The remaining free radicals then attack the base oil and break it down. This gives rise to the condensation stage where the viscosity of the lubricant increases and the by-product becomes insoluble.

HOW CAN OXIDATION BE IDENTIFIED?

Typically, during oxidation, oxygen gets added to the base oil to form aldehydes, ketones, hydroperoxides, and carboxylic acids. It is important to note the presence of these products as they can help us during our investigation. Their presence can confirm the degradation mode or aid us in determining the next steps in preventing this mode.

Usually, when oxidation occurs, there is the formation of deposits in the components. These can be identified as varnish or sludge and can have many variations which include the products listed above. As such, it is always advisable to have the deposits analyzed by a lab to identify the products present. If we know the products present, then we can define areas of improvement or methods of elimination to prevent oxidation.

Once the lubricant experiences oxidation, it undergoes a couple of changes to its structure and characteristics. As described above, its base oil is broken down by free radicals. Additionally, its viscosity increases due to the condensation stage. The lubricant will also undergo a loss of antifoaming properties and additive depletion (namely antioxidants). These are some characteristics which are common to lubricants that have undergone oxidation.

WHAT LAB TESTS CAN BE PERFORMED?

There are six main tests to determine the presence of oxidation. These tests are usually done as secondary tests as some of them are more expensive than the primary tests. Primary tests are the basic tests such as viscosity (ASTM D445), presence of fuel/water, total acid number or total base number (TAN or TBN), and the concentration of additives, metals, or contaminants. These trigger tests and the secondary tests are covered in more detail in *Lubrication Degradation Mechanisms, A Complete Guide* by Sanya Mathura.

The main tests to assist in determining the presence of oxidation include:

- An increase in acid number – indicates the presence of acids resulting from the oxidation process.
- Rapid color changes – darkening of the oil due to the deposits being present.
- Fourier Transform Infrared (FTIR) test– for the presence of insolubles formed from the oxidation reaction.
- Membrane Patch Calorimetry (MPC) levels outside of the normal range (above 20).
- Rotating Pressure Vessel Oxidation Test (RPVOT) levels less than 25% as compared to new oil.
- Remaining Useful Life Evaluation Routine (RULER) levels less than 25% as compared to new oil. This value represents the level of antioxidants in the oil. Hence, low levels indicate that the antioxidants are decreasing possibly due to oxidation.

HOW CAN OXIDATION BE PREVENTED?

Essentially, one of the best ways to prevent oxidation is to discontinue it at either of the stages before termination. We can either reduce the free radical content or ensure there isn't anything to react with the free radicals to promote the degradation of the lubricant. Another option is to guarantee the antioxidant levels do not get depleted to the point where the base oil is attacked.

One company, Fluitec, has developed an oxidation treatment solution. Through the work of their team of scientists, they have developed a product called DECON AO™ made with Solvancer®. This product essentially keeps the deposits in solution. As such, they do not come out of solution to adhere to the surfaces of the components. Additionally, it keeps the antioxidants at acceptable levels to prevent the breakdown of the base oil.

When coupled with their Electrophysical Separation Process (ESP) system, this allows any varnish which is in the solution to be removed from the oil. The presence of varnish in the oil results in a higher possibility of creating more varnish. Thus, if the varnish in the solution is removed, the existence of varnish is eliminated. Therefore, by combining the two technologies, oxidation can be prevented through the removal of the deposits and rebalancing of the antioxidants. At the time of writing, we are only aware of the capability of this technology from Fluitec. We encourage readers to research other options that may exist in the market after this book's publication.

APPLYING RCA TO A LUBRICATION DEGRADATION-RELATED FAILURE

Finally, we are going to put all the pieces of the puzzle we've learned so far and apply them to basic lubrication degradation RCA templates in the coming chapters (Chapters 4–9). These templates will be like troubleshooting flow diagrams for specific lubrication degradation mechanisms. Since we have now learned the visual language of the logic tree, it should be a breeze now to read the story being told in this case. Let's get to it!!

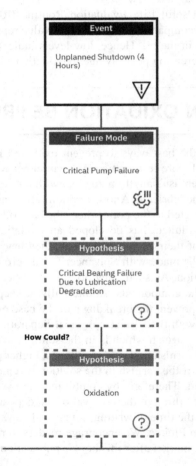

FIGURE 4.1   Lubrication degradation case – oxidation (1).

Our mythical event in this case (the reason we care) is that we suffered four hours of unplanned downtime resulting in a production loss of 240,000 USD ($1k/minute of downtime losses) as shown in Figure 4.1. The facts in the top box support this was due to the failure of a critical pump bearing. From the "crime scene" (facts) it was proven the bearing failed due to lubrication degradation-related issues.

Question:

How could the bearing have failed due to lubrication degradation-related issues?

The potential hypotheses are outlined in Figure 4.2. Now the task is for the RCA team to conduct appropriate tests to determine which of the hypotheses are true and which are not.

Let's start to build our VL (verification log) now. This is usually in a table format which will help us keep track of our evidence/verifications as we go. It will also be a key part of our strategy when we present our findings to leadership. The maintaining of this log is what will make our logic tree stand tall because we have the confidence that it is comprehensive and accurate. Since this is more of an RCA template (versus a case), we have altered the verification log table to provide more commentary to the reader. In Chapter 10, we will analyze an actual case and show the verification log would be filled out in the field. For this and the upcoming examples, the hypotheses will be denoted by [H] and the Verification Technique and Comments denoted by [VTC].

[H] Presence of oxygen and temperature in excess of operating temperature

> *[VTC]* For oxidation to occur, there must be the presence of oxygen and a rise in temperature. Typically, for every 10°C rise above 60°C, the life of the oil is halved. Therefore, one can expect that oxidation may start occurring above 70°C.
>
> Some systems may exceed this temperature during operations; hence, it is critical to understand your normal operating temperatures and let this form the baseline before comparing temperatures.
>
> To verify a rise in temperature, one can review the temperatures on the system log.

[H] LTA Presence of antioxidants

> *[VTC]* As mentioned earlier, antioxidants deplete when they interact with free radicals. Therefore, by monitoring the concentration of the antioxidants, we can estimate the rate at which they are being depleted.
>
> However, if the RPVOT and RULER tests produce results less than 25%, then our base oil will become exposed to the free radicals and the oxidation process can begin.

Now we will explore "How Could" these hypotheses occur.

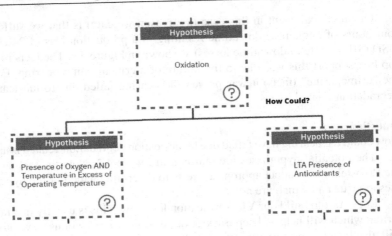

FIGURE 4.2 Lubrication degradation case – oxidation (2).

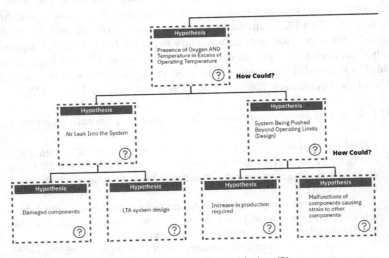

FIGURE 4.3 Lubrication degradation case – oxidation (3).

In our updated logic tree in Figure 4.3, we can see that our next question is "How can we have 'Air leakage Into the System?' and 'System Being Pushed Beyond Operating Limits (Design)'?" As we can see, as we read our story that is unfolding, the two hypotheses proposed by the team are:

1. Air Leak into the System
   a. Damaged Components
   b. Less-Than-Adequate (LTA) System Design

2. System Being Pushed Beyond Operating Limits (Design)
   a. Increase in production required
   b. Malfunctions of components causing strain on other components

These are added to our verification log and must be validated as per below.

[H] (1) Air leak into the system

> *[VTC]* If air leaks into the system, then this could be the source of oxygen, which triggers oxidation. We need to verify the presence of any air leaks into the system.
>
> This can be done by visual inspection of the system or pressurized testing of the system.

[H] (1a) Damaged components

> *[VTC]* One of the ways by which an air leak can occur would be due to a damaged component. Should the component's integrity be violated, this can cause it to allow air to leak into the system.
>
> A visual inspection of the component or pressurized testing can detect any air leaks.

[H] (1b) LTA system design

> *[VTC]* Air leaks can also occur if the system is not designed properly. In that case, there could be areas which allow air to enter the system.
>
> This can be verified through the Original Equipment Manufacturer (OEM), who would have this system designed to be deployed in other applications. Typically, OEMs will also have an Failure Mode and Effects Analysis (FMEA) chart developed during their testing of the system.

[H] (2) System being pushed beyond operating limits (design)

> *[VTC]* If the system is being pushed beyond its design limits, then the temperatures of the overall system may increase. This can be responsible for the increase in temperature throughout the system, which can give rise to oxidation.
>
> To verify whether the system is being pushed beyond its operating limits, we can contact the OEM or check the operating manuals. Typically, these are provided in the operating guidelines. If these guidelines are not documented, then the OEM can be contacted as they are required to document the performance limits of systems during their testing phases.

[H] (2a) Increase in production required

> *[VTC]* Occasionally, an increase in production may be required by the facility to fulfill a particular order. In these instances, the system can be pushed above its operating limits.

This can be verified through an examination of past production levels versus the current values. Afterward, these values can be compared to those given by the OEM to determine whether the system was pushed beyond its limits.

[H] (2b) Malfunctions of components causing strain to other components

[VTC] Typically, components have been designed for particular tolerances. If the system is being pushed beyond its operating limits, this can cause other components to malfunction. As a result, these malfunctions may cause a decrease in the efficiency of the system. This can lead to other components experiencing abnormal loads.

Visual inspections and vibration analysis can assist in determining the malfunction of some components.

Since this is an exploratory logic tree and will serve as a troubleshooting flow diagram, we will consider all possibilities as "True" for now and continue drilling down. Keep in mind the purpose of this final tree will be for the analysts to use this logic tree as a template to follow the hypotheses in their own cases and determine what is "True" and "Not True".

You know the routine by now, and let's update our verification log with how we verified our new hypotheses shown in Figure 4.4.

The following gives us an idea of the verification techniques that can be used when examining whether there was LTA presence of antioxidants. The main hypothesis is the presence of free radicals, which arise due to chemical reactions. We explore the ways in which chemical reactions could produce free radicals and get the following hypotheses:

(1a) Contamination of lubricant introduces catalysts for chemical reactions.

(1b) Adverse operating conditions give rise to chemical reactions.

These are further explored below.

[H] Presence of free radicals

[VTC] The presence of free radicals can impact the presence of antioxidants by depleting them so that they become LTA. Free radicals, as defined earlier, are very reactive, and one method of discovering their presence is using an Electron Spin Resonance (ESR) Spectrometer.

[H] Chemical reactions produce free radicals

[VTC] Free radicals are usually produced during chemical reactions. As such, their presence increases when chemical reactions are ongoing. Since

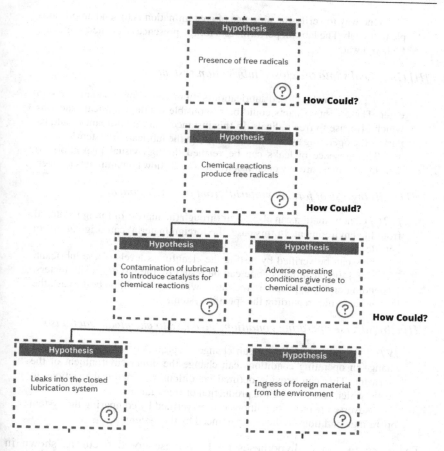

FIGURE 4.4 Lubrication degradation case – oxidation (4).

the production of free radicals can occur with numerous chemical reactions, it is difficult to determine which chemical reaction should be tested. However, when chemical reactions occur, there is usually a change in temperature. Therefore, we should try to obtain the temperature information to help us narrow down on the type of chemical reactions that may be at play.

[H] (1a) Contamination of lubricant introduces catalysts for chemical reactions

[VTC] If chemical reactions are occurring within the lubricant (which are not part of the normal operation) then these can be caused by contamination. If another substance enters the lubricant and begins to cause reactions or acts as a catalyst, then this can produce free radicals.

One way to verify the presence of contamination is to send an oil sample to the lab. The lab can perform tests for the presence of elements, metals, fuel, and water.

[H] (1a.i) Leaks into the closed lubrication system

[VTC] Leaks in the closed lubrication system can allow contaminants to enter. These contaminants could be responsible for the chemical reactions which give rise to free radicals. In this instance, the contaminants could be part of the operating system or run alongside the lubrication system.

The presence of leaks can be verified through visual inspections to determine if there are any openings which would allow contaminants to enter.

[H] (1a.ii) Ingress of foreign material from the environment

[VTC] Contamination can also occur through the ingress of foreign material from the environment. In this case, the foreign material can originate from outside the system.

This can be verified by testing the cleanliness levels of the lubricant. The ISO 4406 cleanliness code can be tested by the lab or on-site meters. Depending on the application, the rating will vary. It will be best to ask the lab for guidelines regarding the specific system.

[H] (1b) Adverse operating conditions give rise to chemical reactions

[VTC] Operating conditions can change very quickly within a system. A change in operating conditions can change the direct environment of the lubricant. This can provide optimal conditions for chemical reactions to occur, which can promote the production of free radicals.

Adverse operating conditions can be verified by comparing the system operating conditions to those experienced by the system.

There are two main hypotheses on how these could occur as shown in Figure 4.5. Either damaged components or seals could attribute to the leak in the closed system or the system had a LTA design. These will now be explored in detail below.

[H] Leaks into the closed lubrication system

[VTC] These were covered above.

[H] (1a) Damaged components or seals

[VTC] If components are damaged, then these can become a source of entrance for contaminants. Typically, most closed systems have seals which ensure that no lubricant escapes and prevents anything from entering the system. Should these become damaged, then leaks can exist within the system.

A visual inspection can determine whether the seals or components are leaking.

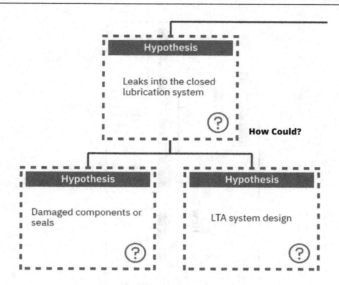

FIGURE 4.5   Lubrication degradation case (5).

[H] (1b) LTA system design

> *[VTC]* Should the system be designed in such a manner to allow leaks into the closed system then this would be LTA. This LTA system design could allow leaks into the system, which will promote contamination, the production of free radicals, and oxidation.
>
> The OEM can be contacted about the system design as they would have this system in other applications and can verify if all the systems are experiencing the same leakage issue.

The above explores the different ways in which ingress of foreign materials can occur from the environment. We must remember that these foreign materials may have been responsible for the contamination of the lubricant, which introduced chemical reactions or catalysts. These chemical reactions, in turn, produced free radicals, which were responsible for the LTA antioxidants. Consequently, this gave rise to oxidation.

There are three main hypotheses when exploring the ingress of foreign materials from the environment, as shown in Figure 4.6:

1. Openings allowing the material to enter the system
2. Wrong lubricant placed into the system
3. Contaminated lubricant placed into the system

Each of these hypotheses will now be explored in greater detail.

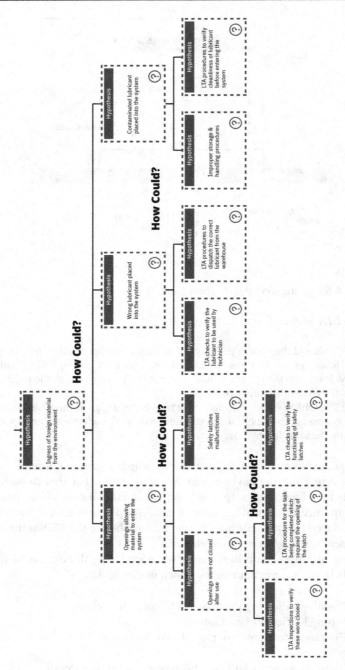

FIGURE 4.6 Lubrication degradation case – oxidation (6).

[H] Ingress of foreign material from the environment

[VTC] As mentioned in the discussion above, contamination can also occur through the ingress of foreign material from the environment. In this case, the foreign material can originate from outside the system.

This can be verified by testing the cleanliness levels of the lubricant. The ISO 4406 cleanliness code can be tested by the lab or on-site meters. Depending on the application, the rating will vary. It will be best to ask the lab for guidelines regarding the specific system.

[H] (1) Openings allowing material to enter the system

[VTC] Foreign materials from the outside can enter the system if there are any openings. A visual inspection can easily detect whether an opening has remained open when it should be closed.

[H] (1a) Openings were not closed after use

[VTC] One way of openings allowing the material into the system is when they were not closed after use. Quite frequently, hatches or other openings are opened to perform maintenance and then left open. Once these are left open, then foreign material can enter the system. A visual inspection can verify whether the opening has remained open after use.

[H] (1a.i) LTA inspections to verify these were closed

[VTC] One such way which an opening could not be closed after use could be LTA inspections. Inspections usually occur on equipment before they are started, while they are running, and before they are shut down. During maintenance, the equipment could be in any of these phases hence inspections should continue to ensure that the openings are closed.

The inspection checklist can be examined to ensure whether this forms part of the listing.

[H] (1a.ii) LTA procedure for the task being completed which required the opening of the hatch

[VTC] Another way in which an opening could not be closed after use is if the procedure did not specify it. Typically, when an opening is opened, this is done for some maintenance procedure to be carried out. For each maintenance procedure, there should be an accompanying task list which identifies all elements of the procedure being performed. However, if the procedure is LTA, then the person completing the task may not close the opening.

The procedure involving all maintenance activities for this opening should be examined and amended if necessary.

[H] (1b) Safety latches malfunctioned

[VTC] If the safety latches on the opening malfunctioned, then they could allow the material to enter the system. This can be verified with a physical inspection.

[H] (1b.i) LTA checks to verify the functioning of safety latches

[VTC] The safety latches could malfunction if LTA checks were performed. On the inspection checklist, there is typically a task assigned to verify the functioning of latches on openings. If this task was not completed, then the functioning of the safety latch could not be verified.

Completed inspection checklists should be reviewed to ensure whether the task was completed or if the checklist was completed hurriedly.

[H] (2) Wrong lubricant placed into the system

[VTC] Another source of foreign material into the system can be the use of the wrong lubricant. During the top-up of the sump, maintenance or operations could mistakenly place the wrong lubricant in the system. This could cause the ingress of foreign material, leading to contamination, which can produce chemical reactions. Consequently, this can produce free radicals which can lead to LTA presence of antioxidants and oxidation.

Oil samples can be sent to the lab to verify the type and grade of lubricant in the system.

[H] (2a) LTA checks to verify the lubricant to be used by technician

[VTC] Sometimes, technicians or personnel assigned to do top-ups or changes may not take the time to verify that they are placing the correct lubricant into the equipment.

One way to verify that they did put the correct lubricant into the sump would be to confirm the lubricant used from dispatching or perform an oil analysis to determine if the lubricant meets the specifications that are consistent with its nature.

[H] (2b) LTA procedures to dispatch the correct lubricant from the warehouse

[VTC] If the warehouse or dispatching unit was not aware of the correct lubricant to be dispatched, then the wrong lubricant can enter the system. Some items to explore would be proper labeling of the lubricant, location of the lubricant (all gear oils in one place, sorted by viscosity, for example), and proper descriptions of the type of lubricant that match the equipment in which it will be used.

One can examine the procedure involved in dispatching lubricants to confirm whether the error occurred in this area or not.

[H] (3) Contaminated lubricant placed into the system

[VTC] Foreign material can also get into the system if contaminated lubricant was placed in the system. This can lead to chemical reactions occurring to produce free radicals which lead to LTA antioxidants and eventual oxidation.

An oil sample can be carried to the lab to verify the cleanliness and to detect if there are any contaminants in the current oil.

[H] (3a) Improper storage and handling procedures

[VTC] Lubricant can become contaminated if there are improper storage and handling procedures. If the lubricant is being dispensed into a smaller container, this container can contain contaminants, which will be transferred to the oil. Similarly, the dispensing equipment could also contain contaminants, which can be transferred to the oil. Additionally, improper storage of the lubricant (exposed to the elements, non-ideal temperatures, and leaving lubricant open in the atmosphere) can introduce contaminants.

An audit should be conducted to check the storage and handling procedures of the lubricants in use.

[H] (3b) LTA procedures to verify cleanliness of lubricant before entering the system

[VTC] Contaminants can also be introduced if the cleanliness of the lubricant is not verified before it enters the system. In some applications, manufacturers recommend a particular ISO/NAS cleanliness rating before placing the lubricant in the system. If these are not observed, then contaminated lubricant can enter the system.

One can examine the procedures involved in topping up or replacing the lubricant into the sump to verify if the cleanliness standard is observed.

As we concluded earlier, there are two hypotheses which can lead to oxidation. Either there is the presence of oxygen and temperature in excess of the operating temperature or there is LTA presence of antioxidants. If there is LTA presence of antioxidants, then we can have two hypotheses: either there is a presence of free radicals or the lubricant can have LTA lubricant specifications. We have already explored the hypothesis of the presence of free radicals. We will now explore the hypothesis of LTA lubricant specifications.

Above, we examined the other hypothesis of "How could a lubricant have LTA presence of antioxidants?" This hypothesis of "LTA lubricant specifications" leads us to two main hypotheses:

1. Lubricant was not blended properly.
2. LTA antioxidant levels were not appropriate to protect the lubricant.

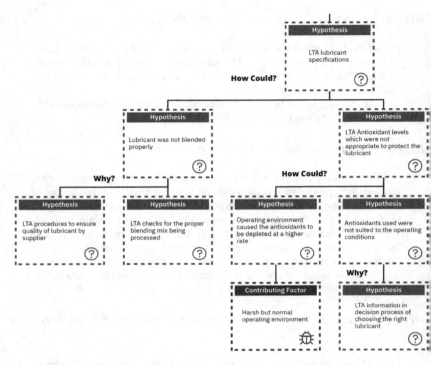

FIGURE 4.7 Lubrication degradation case – oxidation (7).

We will explore these below as shown in Figure 4.7.

[H] LTA lubricant specifications

[VTC] A lubricant can have LTA presence of antioxidants if the lubricant specifications have not been met. In these cases, the lubricant does not meet the criteria required to perform its functions.

This can be verified by matching the lubricant specifications with those of the OEM for that particular component.

[H] (1) Lubricant was not blended properly

[VTC] Errors can occur within the blending plant, which may cause the lubricant to not meet the actual specifications provided by the supplier. If the lubricant was not blended properly then issues can occur where the antioxidants (or other additives) are not allowed to perform at their expected capacity. This can be verified through a certificate of analysis which is usually done by the blending plant for each batch of lubricant blended.

[H] (1a) LTA procedures to ensure the quality of lubricant by supplier

[VTC] The quality of the lubricant is directly affected by its blending process. If the lubricant is blended with low-quality products or the products have not been blended properly then the quality of the lubricant will be affected.

A review of the procedures regarding the final quality of the lubricant and the quality of the products that were used should be conducted.

[H] (1b) LTA checks for the proper blending mix being processed

[VTC] If the quantity or type of products being blended did not match the exact recipe of the blend, then it would not blend properly. For instance, if another product were used erroneously, then it would not have the same compatibility as the one that should have been used. Similarly, if the quantity of the product has been altered then the right balance would not be achieved.

This is like baking a cake; if we use the wrong type of flour or we add too much or too little, the property of the finished product is affected.

A review of the materials consumed for the blend can be matched to the actual recipe of the blend to ensure the correct blend was completed.

[H] (2) LTA antioxidant levels which were not appropriate to protect the lubricant

[VTC] Antioxidants are responsible for normalizing the free radicals. Their role, as described earlier, is sacrificial. Once they have depleted, then the free radicals will attack the base oil and oxidation will occur. Should the levels of the antioxidants be inappropriate, then they will not be able to protect the base oil.

An oil sample can be sent to the lab for the RULER test to be completed. If the results are less than 25% of the original value (100%) then the levels are LTA.

[H] (2a) Operating environment caused the antioxidants to be depleted at a higher rate

[VTC] If the lubricant is in an environment which promotes oxidation (or the production of free radicals) then the levels of the antioxidants can deplete rapidly.

Typically, the standard operating environment is outlined by the OEM in their guidelines. These should be reviewed and compared with the actual operating conditions.

[H] (2a.i) Harsh but normal operating environment

[VTC] Sometimes, the environment of the lubricant is harsh, and this is considered normal. One instance can be in an ammonia compressor where ammonia may come in contact with the lubricant. In these cases, the OEM should be consulted to know the limits of the operating environment and whether the system and lubricant can withstand these.

[H] (2b) Antioxidants used were not suited to the operating conditions

[VTC] There are different types of antioxidants (primary and secondary), and these are selected as per the base oil type (assigned API group) and application. In some instances, not enough research is completed and the antioxidants chosen may not be ideally suited for the operating conditions.

Verification of the type of antioxidants and base oil can be derived from the lube supplier. Additionally, the suitability of the antioxidants for the environment in question can also be gained from them.

[H] (2b.i) LTA information in decision process of choosing the right lubricant

[VTC] Many oils exist in the market. Some of these are generalized (such as turbine oils), whereas others are more specific (turbine oils in ammonia applications). However, the decision of selecting the correct lubricant must be supported by documents.

The process for the selection of the lubricant should be reviewed to ensure that it was adequate and allowed for the most suitable lubricant to be chosen.

As revealed by this tree, there are many physical, human, and latent roots involved in the investigation of oxidation as a lubricant degradation mechanism. Typically, within the industrial sector, a larger focus is placed on finding the physical root, adjusting this, and hoping oxidation does not occur again. We need to be able to take a deeper look into the real root causes of these degradation mechanisms and ensure these are addressed. It is only when the real root causes are addressed, then can the challenge of oxidation be overcome.

In the upcoming chapters (5–9), we will be providing mini-troubleshooting flow diagrams for each of the Lubrication Degradation Failure Mechanisms. They are generic in format to explore many possibilities, as opposed to any specifics if we were to only follow a single case study with unique variables at play. In Chapter 10, we will wrap up with how to apply these mini-troubleshooting flow diagrams to an actual case study.

With the format we've opted to use, we highly encourage analysts reading this book to continue to drill down the nodes where we dropped off, by continuing to ask "Why". When the solutions become obvious, it is our rule of thumb to stop drilling and start developing and implementing corrective actions!

Building a Tree for the LDM: Thermal Degradation

5

WHAT IS THERMAL DEGRADATION?

As the name indicates, thermal degradation involves the degradation of the lubricant due to excessive heat. In this case, as temperatures exceed 200°C, the lubricant can become cracked. When we talk about the cracking of a lubricant, we actually mean the thermal stability of the lubricant has been exceeded to an extent which shears the molecules. This leads to a decrease in viscosity.

During the shearing process, small molecules become cleaved off. These small molecules can either volatize or condense. If they volatize, they do not leave a deposit. On the other hand, if they condense, dehydrogenation takes place. Coke is formed as the final deposit but there are several variations until the final deposit. Thus, carbonaceous deposits and lacquer are the usual results of thermal degradation.

We can think of a block of ice being heated to the point that it turns into water. Some of the water molecules may evaporate but the others will turn into a liquid, which we will have to clean up. Unlike the block of ice, once the lubricant is thermally degraded, there is currently no method for quickly reformulating the lubricant. Typically, in these cases, an oil change would be required.

DOI: 10.1201/9781003252030-5

WHAT LAB TESTS CAN BE PERFORMED?

The lab tests associated with thermal degradation are:

- Viscosity – when the lubricant becomes sheared, it experiences a decrease in its viscosity.
- Color – due to the carbonaceous deposits formed within the lubricant, the color will rapidly darken.
- FTIR – the condensed molecules which were cleaved off undergo dehydrogenation. During this process, carbonaceous deposits are formed with coke being the final deposit. FTIR can identify the presence of these deposits.

WHAT CAN BE DONE TO PREVENT THERMAL DEGRADATION?

The main factor responsible for thermal degradation is the presence of high temperatures, which affects the thermal stability of the lubricant. Hence, to prevent it, we must find ways of reducing the heat in the system. One way of doing this is to increase the clearances in the components. If the clearances are larger, it can allow for more fluid to flow and heat to be dissipated at a faster rate. OEM approvals will be required before its implementation.

Another way of reducing temperatures is to increase the residence time of the oil in the main sump. This allows the lubricant more time to dissipate the accumulated heat from the system. To increase the residence time, a simple solution can be to enlarge the size of the main sump or have primary and secondary sumps. These both allow the lubricant to spend more time in the sump before it is again circulated through the system.

APPLYING RCA TO A THERMAL DEGRADATION-RELATED FAILURE

Let's investigate some of the reasons behind a thermal degradation failure. This can be used as a troubleshooting diagram for similar types of situations.

Our mythical event (as from the last chapter) is that we suffered an unplanned shutdown. As with any unplanned shutdown, this causes a lot of downtime and increased expenses. For this case, we know that there was a critical pump failure resulting from a bearing failure due to lubrication degradation. Our first hypothesis, as shown in Figure 5.1, is the bearing failed due to thermal degradation.

We must now ask the question "How could the bearing fail due to thermal degradation?" For thermal degradation to occur, the base oil must be heated until it gets to the point of becoming thermally unstable. We must now investigate how could this happen? This then leads us to hypothesize that the temperature must be in excess of 200°C. How could this happen? There are two possible ways: either this temperature is reached as a result of normal operating conditions or there was an overload of a related component. We will explore some verifications below. Remember, the hypotheses are designated by [H] and the Verification Techniques and Comments are denoted by [VTC].

[H] Thermal degradation

[VTC] The presence of carbonaceous deposits confirms that thermal degradation has, in fact, occurred. These deposits can be spotted visually. However, they should be carried to a lab to properly identify the compounds present in them. In thermal degradation, deposits can exist in several variations before becoming completely carbon.

[H] Base oil heated until it becomes thermally unstable and cracks

[VTC] Thermal degradation occurs when the base oil becomes thermally unstable and cracks. This occurs due to excessive heat. The long carbon chain in the base oil gets broken (much like the block of ice being heated in the example above).

We can confirm the presence of an increase in temperature through the temperature logs for the equipment. When the lubricant is cracked, there will be a decrease in viscosity; hence, this can be confirmed by a lab through the viscosity test.

[H] Temperature in excess of 200°C

[VTC] Lubricants can withstand some high temperatures. However, once those temperatures exceed 200°C, thermal degradation can occur. We can confirm the temperatures by reviewing the temperature logs of the equipment or using a temperature gun to determine the actual operating temperatures of the equipment. (If there was a failure and that unit has been turned off then the temperatures of the surrounding equipment can be measured to determine an average operating temperature, should the components not be outfitted with temperature sensors.)

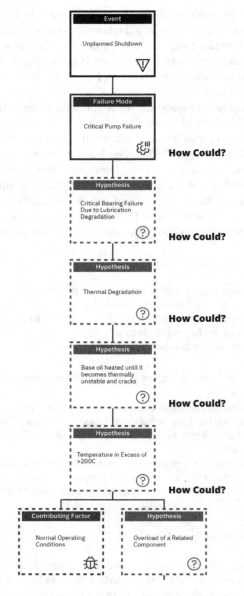

FIGURE 5.1   Lubrication degradation case – thermal degradation (1).

[H] (1a) Normal operating conditions (contributing factor)

[VTC] We have already alluded to the rule of thumb where every 10°C rise above 60°C essentially halves the life of the oil. Equipment manufacturers typically state the operating temperatures for their equipment. This is usually based on the composition of the component material as well as the estimated operation of the entire system.

Operators can verify the system temperatures and compare these to those stated in the equipment manual. There may be instances where the temperature in excess of 200°C is the normal operating temperature. In this case, a lubricant suited for this temperature should be used.

[H] (1b) Overload of a related component

[VTC] A rise in temperature above the normal operating temperature can occur if a component is overloaded. In this instance, the component has been pushed outside of its operating zone. If the component is stressed, then it will require more energy to perform its function, which is translated to an increase in heat and by extension the temperature increases.

Verification of the actual operating temperatures and conditions of the component can be compared against those specified by the OEM.

Let's take a deeper dive into the overload of a related component and how could that occur within our system. There are three hypotheses for this overload:

1. Too much lubricant
2. Not enough lubricant
3. Change in operating conditions

For the ease of graphical reference, we will explore the first hypothesis of "Too much lubricant", as shown in Figure 5.2, and explore it in further detail below.

[H] Too much lubricant

[VTC] One of the most common failure mechanisms for bearings is over greasing! In the past, operators/maintenance personnel would typically pump grease into a port until they saw it coming out on the next side. This is the method they used to ensure that the component received sufficient grease.

In fact, this weakens the seals (or it may pop the seals in some cases) causing the grease to no longer be contained. Thereby, the grease cannot perform its function of lubrication as it leaks out.

Additionally, now with the excessive volume of grease is applied, the components have more work to do while moving. This, therefore, increases the amount of energy that needs to be output thereby increasing the overall temperature of the component.

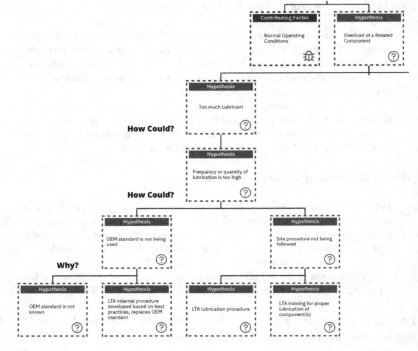

FIGURE 5.2   Lubrication degradation case – thermal degradation (2).

The OEM manuals typically state the quantity of lubricant required for the components. This quantity should be verified with the quantities used by the technicians.

[H] Frequency or quantity of lubrication is too high

[VTC] If the frequency or quantity of lubrication is too high, this can lead to a larger volume of lubricant being applied to the component. With this higher volume of lubricant, churning can occur and can lead to an increase in the stress experienced by the component and an increase in temperature.

The frequency and quantity of lubricant being applied can be compared to the OEM guidelines for this particular application.

[H] (1) OEM standard is not being used

[VTC] The OEM typically suggests the frequency or quantity of lubrication for each of the components. If the OEM standard is not used, then this could cause an abnormal frequency or quantity of lubricant to be used instead.

The OEM manual can be checked against the current practices to verify if it is being followed or not.

[H] (1a) OEM standard is not known

[VTC] There are times when the OEM standard is simply not known. This can occur when the OEM manual has gone missing (for a very old piece of equipment) or has never existed (the OEM did not input this information). In these cases, the OEM can be contacted directly to find out the guidelines to be used.

We can perform checks on the availability of the OEM guidelines to the facility.

[H] (1b) LTA internal procedure developed based on best practices, replaces OEM standard (where LTA = Less Than Adequate)

[VTC] Internal procedures, which have no failure history, can be developed by the facility based on their experience with a similar type of equipment in the past. Hence, these best practices can replace the OEM guidelines. However, this may not always be the most ideally suited procedure as components or applications may have changed compared to the previous best practice cases.

Checks can be performed for the development of the procedures for lubricating the components.

[H] (2) Site procedure not being followed

[VTC] Another possible cause of using too much lubricant in the component could be nonconformance to the site procedure. If the procedure was not followed, then the person executing the procedure could have placed a higher quantity or increased the frequency of lubrication.

This can be verified through the checklists being examined for this component.

[H] (2a) LTA lubrication procedure

[VTC] If the procedure is LTA, then it may not describe the quantity or frequency of application of the lubricant. For instance, some procedures can state, "Grease component" but it does not indicate the exact quantity or frequency. In these cases, the procedure is LTA.

One can examine the current procedure to verify whether it is LTA or not.

[H] (2b) LTA training for proper lubrication of component(s)

[VTC] Another cause for site procedure not being followed is LTA training for the proper lubrication of components. If the personnel responsible for the lubrication of the components have not been trained on how to apply the lubricant, then they will not be equipped to follow the site procedures.

Past training logs can be examined to verify whether or not the personnel has been trained.

As stated earlier, another hypothesis for the overload of the component can be not enough lubricant being applied to the component. Figure 5.3 examines the route that can be taken when investigating this hypothesis. There are three main hypotheses that can be followed when investigating the initial hypothesis of not enough lubricant. These are:

1) The frequency or quantity of lubrication is too low
2) Leak in the component
3) Inappropriately specified lubricant

For the ease of graphical reference, the first hypothesis will be followed in Figure 5.3 and the verification explored below.

[H] Not enough lubricant

[VTC] A component can become overloaded if there is not enough lubricant. In the absence or reduced quantity of lubricant, the lubricant cannot perform its intended function in reducing friction. As such, greater force is required for the component to be moved which can lead to overloading.

Visual checks can be made to verify the levels of lubricant in the component.

[H] Frequency or quantity of lubrication is too low

[VTC] If the frequency or quantity of lubrication is too low, then there will not be enough lubricant to prevent the buildup of friction and by extension prevent overloading.

A review of the maintenance checklists can be done to verify the frequency or quantity of lubrication.

[H] (1) Internal procedure not being followed

[VTC] If the internal procedure is not being followed, then the required frequency or quantity of lubricant is not being dispensed.

Interviews with the staff responsible for lubricating the components can determine whether the procedure is being followed or not. Another verification method can be to perform physical inspections of the equipment before and after maintenance routes have been completed.

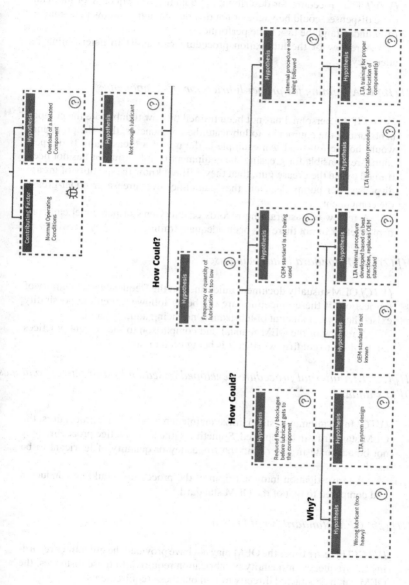

FIGURE 5.3 Lubrication degradation case – thermal degradation (3).

[H] (1a) LTA lubrication procedure

[VTC] The procedure for describing the quantity or frequency of lubricant to be dispensed could be vague or not detailed enough to allow personnel to fully understand the task to be performed.

A review of the lubrication procedure can assist in determining its adequacy.

[H] (1b) LTA training for proper lubrication of component(s)

[VTC] If the personnel has not been trained on how to lubricate the components or use the equipment to lubricate the components, then the procedure would not be followed. An example is the use of a grease gun. If the personnel responsible for greasing the equipment is unfamiliar or has not been trained to use the grease gun, then they will not know the quantity of grease dispensed per pump. As such, they can either over-grease or under-grease the component.

A review of past training records or interviews of personnel can confirm whether or not there has been adequate training.

[H] (2) OEM standard is not being used

[VTC] OEMs usually document guidelines on the frequency and quantity of lubrication. If these guidelines are not being followed, then the possibility exists the correct amount of lubricant is not being administered.

A review of the OEM manual and comparison to the current practices can be done to confirm whether it is being used or not.

[H] (2a) LTA internal procedure developed based on best practices, replac OEM standard

[VTC] In an attempt to standardize maintenance routes and procedures, the OEM standard can be replaced. Sometimes, the best practice procedure may not be adequate in describing the frequency or quantity of lubricant to be used.

An investigation into the origin of the procedure should be conducted and compared to that of the OEM standard.

[H] (2b) OEM standard is not known

[VTC] There are times the OEM may not have provided the guidelines regarding the frequency or quantity of lubrication required. In these instances, the OEM can be contacted directly to find out these requirements.

A review of the OEM manual can ascertain if any guidelines have been provided regarding the frequency or quantity of lubricant for the component.

[H] (3) Reduced flow/blockages before lubricant gets to the component

[VTC] Another way in which there can be too little lubricant is if there is something preventing the lubricant from getting to the component. Should there be a reduced flow or a blockage, then an adequate amount of lubricant will not get to the component.

A physical inspection of the surrounding components or the lubrication system can confirm the presence of blockages while an ultrasonic flow meter can detect changes in the flow rate of the liquid.

[H] (3a) LTA system design

[VTC] If the system is not designed to accommodate an adequate flow of the lubricant, then this can hamper the amount of lubricant that reaches the component. System limits can be manipulated depending on the application and this can impact the flow of the lubricant.

If a component has been substituted into the system, then this can also impact the flow rate if the component does not meet the system specifications.

The OEM can be consulted to verify the system is meeting the required lubricant flow rate.

[H] (3b) Wrong lubricant (too heavy)

[VTC] The rate of flow of a lubricant is heavily dependent on its viscosity. The higher the viscosity of a lubricant, the thicker it appears and the slower it moves. If the wrong lubricant is used or a heavier lubricant is being used instead of the OEM recommended viscosity, then this can impact the flow rate causing a reduced amount of lubricant to reach the component.

The viscosity of the lubricant can be compared to the OEM-recommended to ensure that the right viscosity is used.

Let's investigate the hypothesis of a leak in the component being the cause of not enough lubricant, which leads to an overload of a related component. In Figure 5.4, there are only two hypotheses which could lead to a leak in the component:

1) An opening exists, allowing the lubricant to escape outside.
2) An opening allows system products to displace the lubricant.

Let's explore these hypotheses a bit more and find out where they can lead.

[H] Leak in component

[VTC] The quantity of lubricant present can be depleted through a leak in the component. Once the lubricant is allowed to escape then there is no longer an adequate amount to fulfill its function. This can lead to an overload of the

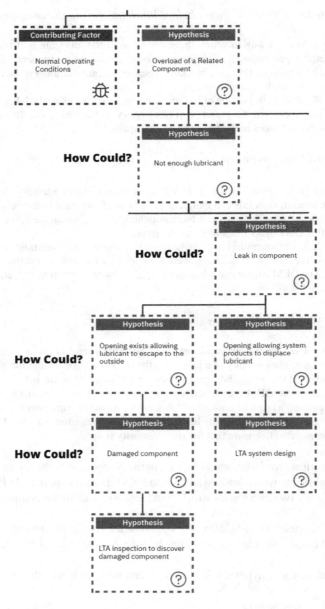

FIGURE 5.4   Lubrication degradation case – thermal degradation (4).

component as it now has a larger load to overcome. This can increase the temperature and by extension lead to thermal degradation.

A physical check of the component can verify whether any leaks exist. Additionally, ultrasonic technology can be used to detect hairline fractures or other leaks.

[H] (1) Opening exists allowing lubricant to escape to the outside

[VTC] Lubricant can escape to the outside of the component leaving little to no lubricant in the actual component. This will impair the function of the component and can lead to overloading. Lubricant can escape through an opening in the component.

A physical check or the use of ultrasonic technology can confirm the presence of a leak.

[H] (1a) Damaged component

[VTC] An opening can occur through a damaged component where the damaged area allows the lubricant to leak to the outside of the component.

A physical inspection can be done to verify the integrity of the component.

[H] (1a.i) LTA inspection to discover damaged component

[VTC] Damaged components may not be observed during inspections, and as a result, the leak can exist and deplete the lubricant from the component.

A review of the inspection list can confirm whether adequate inspection items are mentioned for this area.

[H] (2) Opening allowing system products to displace lubricant

[VTC] Another method of depleting lubricants from their designated area is through displacement. In some instances, an opening can exist which allows another substance to enter the lubricant chamber and displace the lubricant. One example is in an ammonia system where the liquid ammonia may run alongside the lubricant chamber. An opening in the chamber can allow ammonia to enter and displace the lubricant thereby leaving an inadequate amount of lubricant to perform its function.

An oil sample can be sent to the lab from the component. This can confirm the presence of another substance in that particular component through the elemental test. Ultrasonic technology can also be used to detect the leak in the system.

[H] (2a) LTA system design

[VTC] The system could have been designed poorly to allow for cross-contamination of products with the lubricants at particular system conditions.

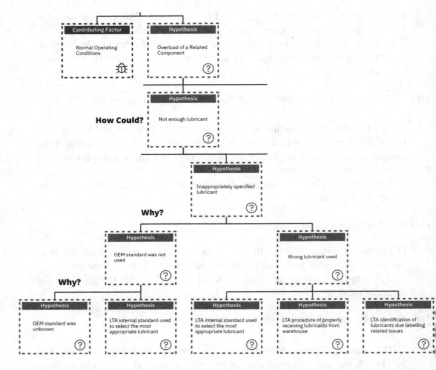

FIGURE 5.5 Lubrication degradation case – thermal degradation (5).

For instance, if the pressure of the system exceeded a certain value, this could cause the flow rate of other substance to increase. As a result of the increased flow rate, the lubricant could be displaced from the system causing the lubricant level to become LTA.

The actual system parameters should be cross-referenced with those of the OEM to verify that they are within the limits. Should the limits not be listed, the OEM could also be contacted to verify these.

An inappropriately specified lubricant can also lead to not enough lubricant being present in the component. In Figure 5.5, we will explore how this could result in thermal degradation. There are two ways in which an inappropriately specified lubricant can be selected either:

1) The OEM standard was not used.
2) The wrong lubricant was used.

Let's explore these in some more detail below.

[H] Inappropriately specified lubricant

[VTC] An inappropriately specified lubricant can cause more damage than good as it is not designed for the system and will, therefore, hamper the efficiency of the component's operation.

An oil sample can be obtained and sent to the lab to verify that the properties of the lubricant in the system match those specified by the OEM.

[H] (1) Wrong lubricant used

[VTC] The wrong lubricant could have been used causing the lubricant in the system to not meet the required specifications.

An oil sample can be sent to the lab to confirm the properties of the lubricant in the system. This can then be cross-examined with the properties of the recommended lubricant for the system by the OEM.

[H] (1a) LTA internal standard used to select the most appropriate lubricant

[VTC] Wrong lubricant could have been used if the standard being used to select the most appropriate lubricant was not adequate. Often, personnel on site are required to standardize or consolidate lubricants based on their specifications. If the person is not knowledgeable enough to properly standardize lubricants or select the most appropriate lubricant then their choice can negatively impact the system.

The procedure used to select lubricants should be reviewed, and its adequacy is confirmed with a subject matter expert (SME).

[H] (1b) LTA procedure of properly receiving lubricants from warehouse

[VTC] If the warehouse did not have an adequate procedure for receiving lubricants, then they could have easily received the wrong lubricant and gave this to the end user. Warehouse personnel should be trained on ensuring the product received from the supplier matches the facility order and this should be in their receipt procedure.

The procedure for receiving lubricants should be reviewed and adjusted accordingly.

[H] (1c) LTA identification of lubricants due to labeling-related issues

[VTC] Another way the wrong lubricant could have been used is if the user did not know it was the wrong lubricant. Sometimes, drums of lubricants may not be labeled properly or perhaps the label is not visible to the user. In these instances, the user can mistakenly select the wrong lubricant. Another common challenge is the decanting of the lubricant into smaller or larger containers. If these are not properly labeled, then the wrong oil can also be mistakenly selected. For instance, if the container is labeled "Hydraulic oil" but the ISO grade isn't specified then it can be easily mistaken.

An inspection of the labeling of lubricants to be used in the field should be conducted to ensure that labels are visible and contain adequate descriptions.

[H] (2) OEM standard was not used

[VTC] Using an inappropriately specified lubricant is an example of the OEM standard not being followed. OEMs stipulate the specifications of the lubricants for their equipment. In some instances, they even provide a list of alternate lubricants which can be used.

The OEM specifications should be matched against those of the lubricant currently in the system to verify whether the standard is followed or not.

[H] (2a) LTA internal standard used to select the most appropriate lubricant

[VTC] Some facilities may use their best practices to select the most appropriate lubricant. However, this may result in an inappropriate lubricant being chosen, which cannot provide the functions for the system.

A review of the procedure for selecting the most appropriate lubricant should be conducted and the OEM's recommendations should form part of the resources for the selection of the lubricant.

[H] (2b) OEM standard was unknown

[VTC] If the OEM standard was not known, then it could not be used. Some OEMs may not list their lubricant specifications in their manual, but they can be contacted about their recommendations for the lubricant to be used.

A quick review of the manual can confirm the presence of these specifications, or reaching out to the OEM is the best alternate option.

Another way in which overload of a component can occur is through a change in operating conditions. In Figure 5.6, we will explore how this change in operating conditions could occur. There are two main hypotheses:

1) Change in system requirements
2) Change in materials being used

Let's explore these a bit further below!

[H] Change in operating conditions

[VTC] Once the operating conditions change, this can directly impact the stresses experienced by the components. In some cases, it can lead to the overload of components.

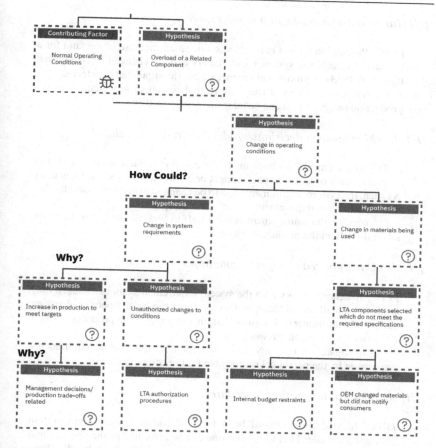

FIGURE 5.6   Lubrication degradation case – thermal degradation (6).

We can verify any changes in operating conditions through interviews with the personnel performing the tasks and the management.

[H] (1) Change in system requirements

[VTC] System requirements are changed at times to make accommodations for changes to the regular operation of the equipment. These may be done to accomplish particular targets.

Investigate the current system requirements and compare these to those observed in the past.

[H] (1a) Increase in production to meet targets

[VTC] Production targets can increase due to an increase in demand for a product. As such, the system requirements may be changed to ensure that these new levels of production are met within the stipulated timeframe.

Interview members of the team to find out about any changes to the production targets. (This can be adapted to your facility.)

[H] (1ai) Management decisions/production trade-offs related

[VTC] Management may have made the decision to increase production to meet a specific order for a special client or certain needs (such as logistical). As such, the system requirements had to be changed to accommodate for this larger production requirement.

Interview the management and the staff to find out about any decisions taken that affect the production rates.

[H] (1b) Unauthorized changes to conditions

[VTC] Another way in which the system requirements can be changed may be due to unauthorized changes. One example can be an operator changing particular requirements to ensure that his team meets the production target or to avoid working on the weekend.

Investigate whether any decisions regarding the system requirements were made by the personnel responsible for the equipment.

[H] (1b.i) LTA authorization procedures

[VTC] Ideally, there should be a level of authorization before system changes are made. This way, everyone can be informed of the changes and consultations can occur to determine if the equipment can handle those changes.

Review any authorization procedures regarding the change of system requirements.

[H] (2) Change in materials being used

[VTC] The composition of materials being used also impact the functionality of components. If the material of the component has changed, then this can affect the system. For instance, if a different type of metal is used to manufacture a pipe, then this new metal may have a higher conductivity of heat compared to the previous pipe metal. This can directly impact the temperatures of the components around the new pipe and the substance being carried.

Investigate for any changes in the material of the components.

[H] (2a) LTA components selected which do not meet the required specifications

[VTC] Sometimes the system requirements are neglected and LTA components are used instead of those meeting the required specifications.

We can verify if the actual component specifications match the required component specifications by checking the manuals of both or contacting the OEM to seek further information. In some cases, the OEM may list alternative components that can be used (which still meet the specifications).

[H] (2a.i) Internal budget restraints

[VTC] Procurement may be under the initiative to reduce costs and could only afford an alternative which did not meet the specifications. This may directly impact the system requirements. Hence, in these cases, when asking for procurement to purchase new components, the requester should state which specifications are non-negotiable. It may cost a lot more when addressing the issue of an unplanned shutdown compared to some dollars off the purchase price.

Interview procurement about the process used to identify and purchase components.

[H] (2a.ii) OEM changed materials but did not notify consumers

[VTC] The OEM may also change the composition of the material of the component and not inform the customer. These changes can occur due to the availability of material or changes within the organization from the OEM's side. While the change may not affect all customers, it could affect customers with specific applications. For instance, if we think about the pipes where the material was changed, this may not affect customers who are transporting substances that do not react with the new material, but it can affect those whose substances can react negatively with the new material.

The OEM can be contacted to ensure that there were no changes to the materials of the components. It would be important to note the serial and batch number of the component when contacting the OEM.

The potential roots outlined by this tree can be used as a guide when understanding the root causes of thermal degradation. It must be noted that we should not stop at the physical or human roots of the issue and hope that this is fixed for the future. We must investigate these incidents thoroughly to ensure that they do not occur in the future.

Building a Tree for the LDM: Microdieseling

6

WHAT IS MICRODIESELING?

Microdieseling is also known as *compressive heating*. It can also be perceived as a form of pressure-induced thermal degradation. In this mechanism, entrained air moves from a low-pressure zone to a high-pressure zone. During the movement, the trapped air produces localized temperatures in excess of 1,000°C. Due to these changes, the outside of the air bubble becomes carbonized and the oil darkens.

A similar situation occurs when we drive at night with our windows up and the air conditioner off. There is a difference in temperatures on the outside of the windshield compared to that on the inside, such as the glass fogs. Similarly, the entrained air bubble's interface becomes carbonized due to the excessively high temperatures that are coming in contact with the oil.

HOW CAN MICRODIESELING BE IDENTIFIED?

Microdieseling can occur within two conditions: it has either a high implosion pressure or a low implosion pressure, both occurring at a low flashpoint. Different results are produced based on the implosion pressure of the entrained air.

If the entrained air has a *high implosion pressure* with a low flashpoint, it can produce ignition products that are incomplete. These can include soot, tars, and sludge. We can liken this to an explosion: it is almost immediate and items around it may get different degrees of burns. On the other hand, if the

DOI: 10.1201/9781003252030-6

entrained air has a *low implosion pressure*, then it experiences adiabatic compressive thermal heating. This will produce varnish from carbon insoluble, which include coke, tars, and resins.

WHAT LAB TESTS CAN BE PERFORMED?

Before lab tests are performed, the components should always be examined by subject matter experts (SMEs). Since microdieseling involves entrained air, we can expect to see cavitation on the surface of the components. This is one of the telltale signs that microdieseling has occurred. Additionally, the lab can confirm this mechanism using Fourier Transform Infrared (FTIR). The FTIR test can identify the presence of the byproducts, which can, in turn, help us understand whether high or low implosion pressures were involved.

WHAT CAN BE DONE TO PREVENT MICRODIESELING?

Entrained air can wreak havoc on the system. It has been introduced into the system from either within the system or outside the system. Should the air enter the system from outside, we need to identify the possible sources of leaks and have these eliminated. One method would be to check for leaks linked to the lube system and ensure these are fixed.

On the other hand, should the air originate from within the system, we need to eliminate these sources. Typically, these can include the return oil line to the sump allowing the oil to flow too quickly or from a higher altitude to introduce air. One way of solving this is to increase the residence time of the oil in the sump, allowing the entrained air to rise to the surface before being cycled through the system. Another option is to include a baffle plate that traps these air bubbles.

APPLYING RCA TO A MICRODIESELING FAILURE

Let's start with the guideline to understanding microdieseling and some ways in which it can occur. This template can be adapted to suit your organization and is intended as a guide to help you through the tough process of getting to the root cause for this mechanism.

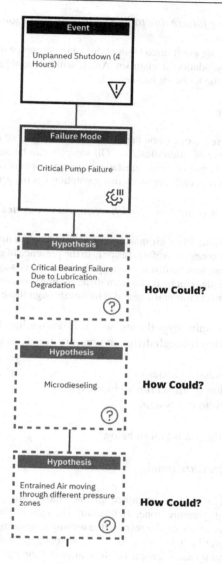

FIGURE 6.1 Lubrication degradation case – microdieseling (1).

The reason that we care (our top box) in this case is the unplanned shutdown of 4 hours, as shown in Figure 6.1. This can be very costly depending on the industry and nature of the application. We ask the question of how could microdieseling occur and then take this line of questioning throughout the tree. Let's start getting those questions answered below!

[H] Critical bearing failure due to lubrication degradation

[VTC] The bearing experienced failure due to the presence of deposits from a lubrication degradation mechanism. An oil sample could be sent to the lab to determine if the lubricant had degraded.

[H] Microdieseling

[VTC] The presence of carbon deposits, coke, soot, tars, or sludge can confirm the presence of microdieseling. Oil samples can be sent to the lab to confirm the presence of these substances and an inspection of the components of the bearing can reveal whether cavitation has occurred or not.

[H] Entrained air moving through different pressure zones

[VTC] In microdieseling, air must be entrained and must move through different pressure zones. Therefore, to confirm the presence of entrained air, we can obtain an oil sample (using a syringe method – this way we can test for the gases). Samples can be taken from different pressure zones to determine whether the air remains entrained as it moves through these zones.

There are three main hypotheses when investigating the reason behind entrained air moving through different pressure zones, as seen in Figure 6.2:

1) Varying pressure zones
2) Introduction of air within a closed system
3) Air leaks into the system

Let's investigate these a bit more below!

[H] (1) Varying pressure zones

[VTC] In microdieseling, entrained air must move through a low-pressure zone to a high-pressure zone. Hence, any changes in the pressure zones where lubricants pass can be responsible for microdieseling.

We can verify the changes in pressure zones by looking at the OEM manuals or the pressure gauges on the equipment through which the lubricant passes.

[H] (1a) Normal operating conditions

[VTC] The movement of air from a low-pressure to a high-pressure zone may be a normal operating condition. This condition can be created when oil

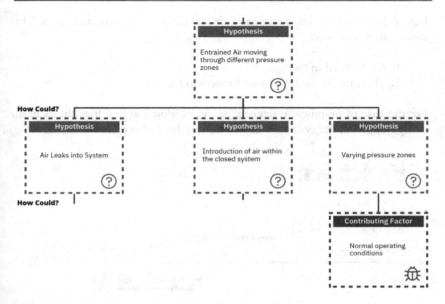

FIGURE 6.2   Lubrication degradation case – microdieseling (2).

enters the suction end of a pump (low-pressure zone) and is then discharged (high-pressure zone). Similarly, this condition can also be created when oil flows from a return line back into the sump. These normal operating conditions may be difficult to change and can be noted as contributing factors.

We can verify these changes in pressure zones by examining the pressure gauges and tracing the movement of the lubricant through the system. These may also be obtained from the OEM manuals.

[H] (2) Introduction of air within the closed system

[VTC] For microdieseling to occur, we must have entrained air. Hence, we need to investigate any sources through which air can be introduced into the closed lubrication system.

We can investigate sources through which air could be introduced in the closed system by reviewing the lubrication system closely and identifying possible areas for the introduction of air.

[H] (3) Air leaks into system

[VTC] Entrained air can be entering the system through a leak; hence, it is coming from outside of the closed system.

We can perform inspections on the components to determine if there are any air leaks into the system.

Let's explore how the introduction of air into a closed system could occur. This can occur in two ways:

1) Air trapped in the system
2) Churning as the lubricant re-enters to the sump

For the ease of graphical reference, we will follow the first hypothesis of air being trapped in the system in Figure 6.3 and further explore these below.

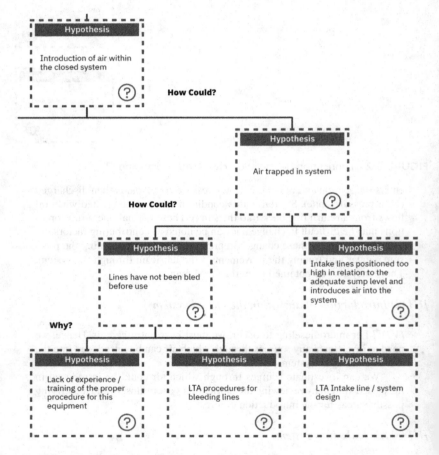

FIGURE 6.3 Lubrication degradation case – microdieseling (3).

[H] Air trapped in system

[VTC] If air is trapped in the system, then it can become entrained and lead to the introduction of air into a closed system which, in turn, leads to microdieseling. While there have been attempts to standardize a test for determining the volume of entrained air, these tests either lacked accuracy or were challenging to perform as per Schrank, Murrenhoff and Stammen (2014).

[H] (1) Lines have not been bled before use

[VTC] If lines carrying lubricant have not been bled before use, then there is a possibility of the entrapment of air in those lines. This is one way through which air can be introduced into the closed system.

This can be verified through interviews of the personnel responsible for filling the lines. We can find out whether or not the lines were bled.

[H] (1a) Lack of experience/training of the proper procedure for this equipment

[VTC] If the personnel filling the equipment were inexperienced in this particular task, then they would not have known about bleeding the lines before use. Interviews can be conducted with the personnel responsible for filling the equipment regarding their past experiences with this task and whether any training was involved for this task.

[H] (1b) LTA procedures for bleeding lines

[VTC] Similarly, if the procedure for filling the equipment did not clearly indicate the task of bleeding the lines, then the task would not have been completed.

A review of the existing procedure can be conducted to identify if the procedure was LTA.

[H] (2) Intake lines positioned too high in relation to the adequate sump level and introduce air into the system

[VTC] If the intake lubrication line is positioned such that the orifice is not entirely covered by the lubricant in the sump, an opening is created through which air can be sucked in. Thus, the air gets introduced into the system.

We can perform an inspection on the intake lines and the levels of lubricant at these locations.

[H] (2a) LTA Intake line/system design

[VTC] If the levels of the sump (in which the intake line is located) fluctuate, then the position of the intake line is extremely important. If the intake line is too high, then when the sump reaches lower levels, air will be sucked into the system. If the position is too low, then any settlements to the bottom of the sump will be forced to recirculate in the system and possibly cause issues.

We can verify whether the system is designed poorly through communication with the OEM and monitoring the levels of the sump to determine if the position is adequate.

Let's take a look at another possibility for the introduction of air within the closed system. In Figure 6.4, we will explore how the churning of lubricant, as it re-enters the sump, could introduce air into the system and give some further detail below.

[H] Churning as lubricant re-enters sump

[VTC] When lubricant re-enters the sump, it can churn introducing air into the lubricant. If the air-release properties of the oil are not adequate, then air can remain entrained and enter the system contributing to microdieseling.
Locations at which churning can be expected should be inspected.

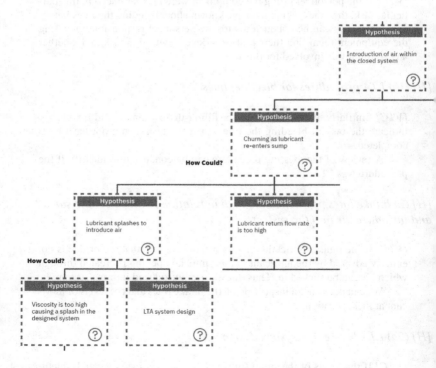

FIGURE 6.4 Lubrication degradation case – microdieseling (4).

[H] (1) Lubricant splashes to introduce air

[VTC] One possibility of introducing air into the lubricant through churning is the splash of lubricant. When a lubricant splashes, we can think of someone performing a cannonball plunge into the pool. Air is introduced into the pool. Similarly, on a much smaller scale, if the lubricant splashes into the larger sump, it can introduce air into the system.

We can verify this by monitoring the areas in which splashes could be expected and introduce baffle plates to reduce the amount of entrained air.

[H] (1a) Viscosity is too high causing a splash in the designed system

[VTC] If the viscosity of the lubricant is very high then this would introduce more air into the system. For comparison, let's think about a coin falling into a pool compared to a larger ball of copper's fall. The larger ball of copper (which is heavier) will undoubtedly introduce more air into the system.

We can verify if the correct viscosity of the lubricant is being used in the system by contacting the OEM.

[H] (1b) LTA system design

[VTC] The system can be designed such that it causes splashing of the lubricant as it re-enters the sump. The manner in which this re-entry occurs can be investigated. We need to identify if this is adequate or not and confirm with the OEM.

Let's continue with the route of air being introduced into the system as the viscosity is too high causing a splash in the designed system, as seen in Figure 6.5. This can lead to two main possibilities:

1) OEM guidelines regarding viscosity were not followed
   a. OEM guidelines regarding viscosity do not exist
   b. The site's best practices were used, replacing the OEM-recommended guidelines
2) Required viscosity was not available and a heavier viscosity substituted
   a. Supplier unable to supply the correct viscosity
   b. Warehousing dispatched the wrong viscosity
   c. The site did not have the available viscosity

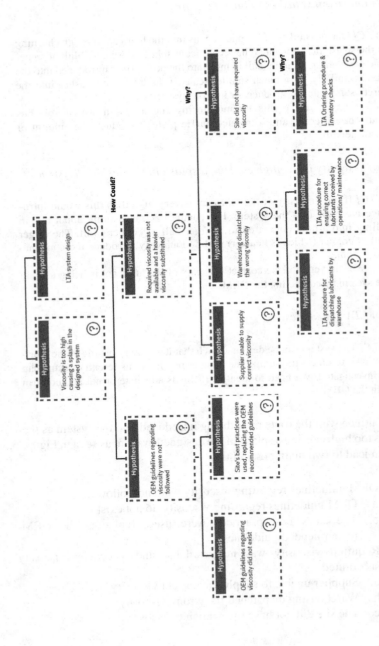

FIGURE 6.5 Lubrication degradation case – microdieseling (5).

Let's explore these in some more detail below.

[H] Viscosity is too high causing a splash in the designed system

[VTC] As discussed earlier, if the viscosity is too high, then it can cause a splash of the lubricant into the system, which can introduce air.

We can verify whether we are using the correct viscosity by consulting the OEM.

[H] (1) OEM guidelines regarding viscosity were not followed

[VTC] The OEM guidelines regarding viscosity could not be followed and this could lead to a higher-than-expected viscosity being used. Thus, the air gets introduced into the system.

We can verify if the OEM guidelines are being followed by matching the viscosity of the lubricant in use to that of the OEM recommendations.

[H] (1a) OEM guidelines regarding viscosity did not exist

[VTC] If OEM guidelines do not exist, then they cannot be followed. Some OEMs may not document the viscosity of lubricant to be used in their systems. In these cases, they may work with another supplier responsible for the lubrication aspect. The component's OEM should be contacted to get the required recommendations.

We can review the OEM guidelines to determine whether or not the relevant guidelines exist.

[H] (1b) Site's best practices were used, replacing the OEM-recommended guidelines

[VTC] Personnel at the site may have decided that the higher viscosity lubricant is more suited for this application due to past experiences. As such, the OEM guidelines are replaced.

This can be verified by matching the OEM guidelines to those that exist at the site.

[H] (2) Required viscosity was not available and a heavier viscosity substituted

[VTC] The required viscosity for the application could have been unavailable and a heavier viscosity is substituted. Sometimes, personnel decides to use a heavier viscosity rather than a lighter one when trying to make a substitution. The correct viscosity is imperative for the application to function at its highest efficiency.

We can verify the availability of the correct viscosity by reviewing the inventory of the warehouse.

[H] (2a) Supplier unable to supply correct viscosity

[VTC] One of the reasons for not having the correct viscosity could be the unavailability at the supplier's end. Due to logistics, blending processes, or simply the ordering process, the supplier could have a depleted stock of the required viscosity.

An interview with the supplier can identify whether this was the issue.

[H] (2b) Site did not have required viscosity

[VTC] Personnel could have used a heavier viscosity if the site did not have the required viscosity readily available. As such, they used whatever they could find! We can verify by checking the inventory levels for that period.

[H] (2b.i) LTA ordering procedure and inventory checks

[VTC] Should there be an issue regarding ordering or relating information on the inventory levels then the site could not have the required viscosity available. When placing orders for lubricants, this is usually done a couple of months in advance and is based on the current inventory and expected orders. Should there be inadequate checks at the warehouse, then an incorrect number can be used to calculate the quantity of lubricants to be ordered.

An audit of the ordering procedure and the variables that are input from the warehouse or other areas can be conducted to ensure that there were no challenges with these.

[H] (2c) Warehousing dispatched the wrong viscosity

[VTC] The warehouse could have dispatched the wrong lubricant and sent over the heavier lubricant instead of the required viscosity. We can verify this through the inventory levels of the lubricants in stock.

[H] (2c.i) LTA procedure for dispatching lubricants by warehouse

[VTC] This can occur if the procedure for dispatching lubricants is vague and does not have the instructions of verifying that the lubricant to be sent is actually dispatched. A review of the dispatching procedure can be conducted.

[H] (2cii) LTA procedure for ensuring correct lubricants received by operations/maintenance

[VTC] If warehousing dispatches the incorrect lubricant, operations or maintenance also have the care of duty to ensure that they are placing the correct lubricant into the equipment. These departments are the last check before the lubricant enters the equipment.

We can review the procedures for receiving lubricants from the warehouse and placing lubricants into the machine. These must include checks to verify that the correct lubricant is being placed in the machine.

We will now explore the remaining two hypotheses of churning as the lubricant re-enters the sump, namely:

1) The return line is at the top of the sump allowing the lubricant to flow at an accelerated rate
   a. LTA system design
2) The lubricant return flow rate is too high
   a. LTA system design
   b. Lubricant flow rate adjusted to compensate for another issue
   c. OEM guidelines not followed on the lubricant flow rate

Let's explore these routes in some more detail in Figure 6.6 and below.

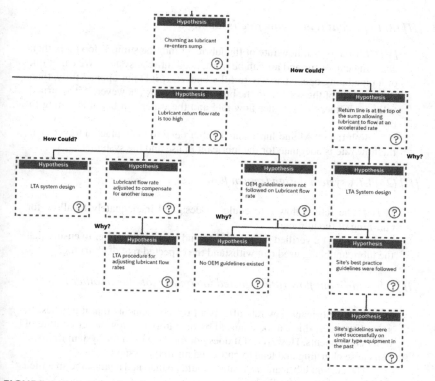

FIGURE 6.6 Lubrication degradation case – microdieseling (6).

[H] (1) Return line is at the top of the sump allowing lubricant to flow at an accelerated rate.

> *[VTC]* If the return line into the sump is at an accelerated height, then the lubricant will experience a drop into the sump. We can think of a coin falling into the pool at the edge of the pool. It would not make much of a ripple. However, if the same coin were dropped from the diving board above the pool, it would create a splash and introduce some air into the system.
>
> We can review the placement of the return line into the sump and verify with the OEM and lubricant supplier whether this height is appropriate.

[H] (1a) LTA system design (return line)

> *[VTC]* The system could be designed poorly or perhaps there were some modifications within that area that caused the return line's placement to be inadequate.
>
> We can verify with the OEM and lubricant supplier regarding the placement of the return line.

[H] (2) Lubricant return flow rate is too high

> *[VTC]* If the return flow rate of the lubricant into the sump is too high, then churning can occur and air can be introduced into the system. We can think of a pipe being opened slowly to fill a bucket. If it is not opened fully, then the flow rate of the water into the bucket will be slow; however, if it is turned fully open, there is a higher flow rate and the water can begin churning in the bucket.
>
> Both the OEM and lubricant supplier need to be contacted to verify if the flow rate is adequate for the lubricant in use and the system.

[H] (2a) LTA system design (return flow rate)

> *[VTC]* The return flow rate could be designed inadequately to allow for churning of the lubricant.
>
> This can be verified with the OEM and lubricant supplier to ensure that the lubricant being used can withstand the expected rate of churning.

[H] (2b) Lubricant flow rate adjusted to compensate for another issue

> *[VTC]* The lubricant flow rate affects all the components that it touches. In some instances, this can be adjusted by operators to allow faster cooling of other components. However, if it does not follow within the system limits, it can cause churning and lead to entrained air in the system.
>
> The current lubricant flow rates should be matched against the specified OEM operating system limits.

[H] (2b.i) LTA procedure for adjusting lubricant flow rates

[VTC] There should be a procedure to adjust the flow rates. This should not be done randomly at the will of the operators or the maintenance personnel.

We can review the procedure involved for the adjustment of flow rates and ensure that these state the limits to be used within the system.

[H] (2c) OEM guidelines were not followed on lubricant flow rate

[VTC] The OEM guidelines would not have been followed regarding the flow rates. If these were not followed, then the system requirements would have changed, and this could allow for churning within the lubricant.

We can review the OEM guidelines and match these to the current flow rates being used.

[H] (2c.i) No OEM guidelines existed

[VTC] If no OEM guidelines existed, then these could not be followed. Typically, all OEMs are required to present optimal operating conditions for their components. However, information may have been omitted from the manual. The OEM should be contacted in this case and the flow rates are verified with those of the system.

We can perform interviews with the team at the facility to find out how the flow rates of the lubricant were derived.

[H] (2c.ii) Site's best practice guidelines were followed

[VTC] The site's best practice for the derivation of flow rates could have been used instead of the OEM guidelines. An interview can be conducted to find out how the flow rates were determined.

[H] (2c.ii.a) Site's guidelines were used successfully on similar type equipment in the past

[VTC] The site may use flow rates from other pieces of equipment to determine the most appropriate flow rate for this system. If this rate was used in the past with no failures, then it would become a best practice flow rate for most equipment within that application.

We can interview the teams to find out how the flow rate was determined for this system.

Entrained air could also enter the system through air leaks. In Figure 6.7, we will examine the route that can be taken if air leaks into the system. Openings along the lubrication system can cause air to enter and be responsible for air

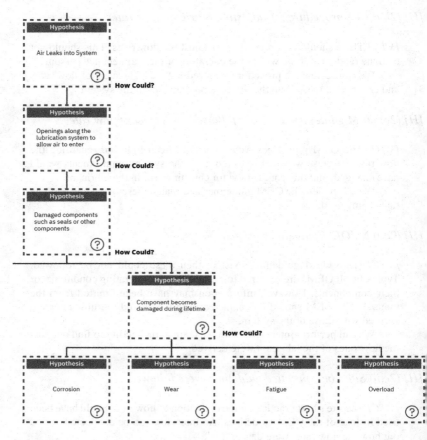

FIGURE 6.7 Lubrication degradation case – microdieseling (7).

leaks into the system. These openings can be a result of damaged components (such as seals or other components). There are two main hypotheses on how these components can become damaged:

1) Defective component
2) The component becomes damaged during its lifetime

For ease of graphical reference, we will focus on the hypothesis of the component becoming damaged during its lifetime and give further explanations on it below.

[H] Openings along the lubrication system to allow air to enter

[VTC] Openings can allow air to enter which, in turn, can be responsible for the air leaks into the system (from the outside). These openings would be responsible for allowing the air to enter the system which can then get entrained.

We can perform inspections to verify the presence of any openings along the lubrication system.

[H] Damaged components such as seals or other components

[VTC] Should any of the components become damaged or mechanically unstable, then this can also allow air to enter the system. Typically, seals are the usual culprits and once their integrity is compromised, air can enter the system. However, if there are other damaged components, these can also be responsible for the entrance of air into the system.

We can perform inspections on the seals and components within the lubrication system.

[H] (1) Component becomes damaged during lifetime

[VTC] The component can become damaged during its lifetime due to wear, fatigue, corrosion, or overload. This would cause its mechanical integrity to become compromised and can allow air into the system.

We can inspect the components to determine if it was damaged due to regular use.

[H] (1a) Wear

[VTC] Wear is the uniform loss of material across a fractured surface. This can be caused by a number of conditions where two surfaces meet and rub resulting in a loss of metal. Fractured components should be properly preserved in a temperature-controlled staging area to prevent further damage. This is until a determination is made by a trained eye (i.e., metallurgist/materials engineer), who will inspect it and/or ship the failed parts to a certified metallurgical lab. This is true for fatigue, corrosion, and overload as well.

[H] (1b) Fatigue

[VTC] Fatigue is a cyclic force on the material that happens gradually over time. This is often evidenced by the observation of "beach marks"-like patterns on a failed surface. As a crack propagates, eventually the remaining material cannot take the load and there will be evidence of an instantaneous fracture zone (IFZ).

[H] (1c) Corrosion

> *[VTC]* This failure pattern is often recognized by "pitting". Corrosion is actually an electrochemical phenomenon. There are many types of corrosion, so such failure mechanisms should be reviewed by trained professionals to determine which type of corrosion is at play.

[H] (1d) Overload

> *[VTC]* Overload is an instantaneous overpowering of the material. A ductile material will deform while a brittle material will fracture. Materials will show a salt and pepper appearance when a brittle fracture takes place. Any material that experiences loading above its ultimate tensile strength (UTS) will fail. Ductile and brittle are the primary types of overload. Again, a trained eye should evaluate and diagnose which failure mechanisms are at play, before drilling deeper to determine the "Whys".

The remaining hypothesis under damaged components is whether the component was defective or not. This leads to two new hypotheses, as shown in Figure 6.8:

1) OEM supplied a defective component
2) Damaged during storage at the site

Let's explore these in some more detail below.

[H] Defective component

> *[VTC]* If a component is defective, then it can be responsible for allowing air to enter the system.
> We can inspect the component to determine whether it was defective (or damaged before it was placed into service).

[H] (1) Supplied by the OEM (retain custody until signed off by client's receiving department)

> *[VTC]* One way in which the component can be defective is if it was supplied in this state by the OEM. There are a number of challenges that the OEM could have faced in this situation ranging from the wrong material being supplied to a malfunction on the manufacturing machines or damaged while it was in their care. We must remember that the OEM has the responsibility for the component until it is supplied to the client. Thus, anything that occurs to the component before being handed over to the client will be the responsibility of the OEM.

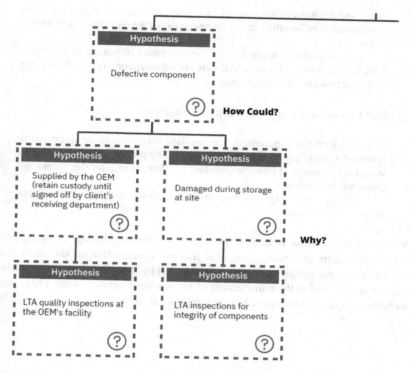

FIGURE 6.8 Lubrication degradation case – microdieseling (8).

We can interview the OEM to find out whether this was an issue faced with that particular batch of components.

[H] (1a) LTA quality inspections at the OEM's facility

[VTC] If the OEM had LTA quality checks, then they would not have noticed the defective component being supplied to their customer.

An interview can be conducted with the OEM regarding their quality checks of their components.

[H] (2) Damaged during storage at site

[VTC] The component could have been damaged on-site (after leaving the custody of the OEM). Sometimes, if components are improperly stored, they can become damaged. For instance, if bearings are stored on a shelf that vibrates every time a piece of equipment is moved, we may introduce fretting before the component has entered its service. The same can be said of the transportation or handling of the components before they reach their installation sites.

We can perform inspections on the equipment in storage to determine whether the current storage conditions are adequate or invite a third-party inspector to audit the storage areas.

[H] (2a) LTA inspections for integrity of components

[VTC] Before the components are installed, they should be inspected to ensure that there are no signs of defects. If they are not inspected, then we would not be aware of possible openings to allow air to enter the system and cause microdieseling.

A review of the type of inspections before the installation of components should be conducted.

While microdieseling is mainly entrained air moving through varying pressure zones, there are many aspects to this mechanism that should be considered when investigating its occurrence. As seen by this tree, there are countless areas that can lead to the entrainment of air in a lubrication system. These need to be investigated thoroughly before conclusions are made.

Building a Tree for the LDM: Electrostatic Spark Discharge

7

WHAT IS ELECTROSTATIC SPARK DISCHARGE?

Electrostatic spark discharge (ESD) is an electrifying degradation mechanism. Static builds up in the oil at a molecular level. This usually occurs when dry oil (oil with no moisture) passes through tight clearances. Typically, this occurs quite often in hydraulic systems as the clearances are very small. Some users have reported hearing crackling noises outside of the equipment indicative of ESD occurring inside.

There are three stages associated with ESD. In the first stage, static electricity builds up in the lubricant to the point where it produces a spark, and then there is a direct increase in temperature above 10,000°C. With this increase in temperature, free radicals are formed, and the lubricant begins to polymerize in the second stage. In the last stage, the lubricant undergoes uncontrolled polymerization, which produces varnish and sludge.

The varnish and sludge produced can either remain in the solution or become deposited. Typically, this results in elevated fluid degradation and the presence of insoluble materials. Usually, users begin noticing a rise in the pressure differentials, and the filters require changing at a higher frequency. With ESD, users can also notice burnt membranes in the filters. These would be the areas which the sparks discharge and burn the membrane.

DOI: 10.1201/9781003252030-7

WHAT LAB TESTS CAN BE PERFORMED?

Although lab tests can be performed to verify the presence of ESD as a lubricant degradation mechanism, the user can also do a visual inspection of the components. In ESD, static usually discharges on the membranes of the filters. Hence, the filter membranes typically have burnt spots indicating the areas in which the spark came in contact. Additionally, users normally hear crackling noises in the equipment when the sparks are discharging.

When ESD has taken place, varnish and sludge are produced. Fourier Transform Infrared (FTIR) testing can help confirm the presence of these materials. Subsequently, gases are also produced during the spark discharge. A dissolved gas analysis (DGA) test to confirm the presence of acetylene, ethylene, and methane also confirms the presence of ESD. Lubricants also experience a decrease in their antioxidants (during the free radical formation before polymerization). As such, a result of less than 25% in the RULER test will also reveal the presence of ESD.

WHAT CAN BE DONE TO PREVENT ESD?

One of the best methods of preventing ESD is using antistatic filters. These filters are designed to remove static from the system preventing its buildup and eventual discharge. There are combinations of using antistatic filters with ESP systems to additionally remove any deposits from the system. This ensures that the oil remains clean and free from static.

APPLYING RCA TO AN ELECTROSTATIC SPARK DISCHARGE FAILURE

ESD is an electrifying degradation mechanism as it involves a lot of sparks! One of the most unique aspects of ESD is the aftermath or telltale signs, which indicate it has occurred. After the buildup of static, it discharges almost always on the filters. Hence, the condition of the filters is very important in identifying this mechanism. Let's explore the guideline below to determine the root cause of ESD.

The reason we care as indicated by the top box is the unplanned shutdown. As seen before in the other templates, we ask the question, "How Could?" as we hypothesize to find the root cause. For this mechanism, let's find out how ESD could occur, as shown in Figure 7.1 and explained below!

[H] Electrostatic spark discharge

> *[VTC]* Operators usually hear crackling outside of the equipment when ESD is occurring inside. While this is not a scientific confirmation (as our versions of crackling can be different), we can take this into consideration as one of the questions we ask personnel during the interview process. As indicated earlier, we can also perform the RULER test (to determine the presence of antioxidants) and dissolved gas analysis for particular gases found in the oil. We can also confirm the presence of ESD with a visual inspection of the filters. If there are patches of the burnt membrane, then we can conclude ESD has occurred.

[H] Static buildup in oil

> *[VTC]* Static builds up in the oil when the dry oil (or oil which does not contain moisture) passes through very tight clearances. This is usually predominant in hydraulic systems where the clearances are particularly smaller than average. Static builds up at a molecular level. ASTM D2624 is the test used to determine the electrical conductivity of aviation and distillate fuels. This test can also be applied to industrial oils to determine their conductivity. The lab can advise on the threshold value for particular oils.

[H] (1) Clearances too tight

> *[VTC]* Should the clearances be too tight or small, the molecules can build up static and lead to ESD. As noted earlier, this is typical of hydraulic systems as they have smaller clearances to allow for the transfer of force through the lubricant.
>
> We can review the clearances of the system against those of the Original Equipment Manufacturer (OEM) and determine if there are any discrepancies.

[H] (2) LTA grounding of system

> *[VTC]* When static is built up within the oil, there must be a path for grounding. Should the system not have adequate paths to allow for grounding, then the static will remain in the oil and can lead to ESD.
>
> A review of the system design can determine whether any grounding mechanisms are present. These can be compared to the OEM suggestions for the system.

[H] (3) LTA conductivity of oil

> *[VTC]* Static must have a medium through which it is conducted. If the level of conductivity in the oil allows for the transmission of static, then this makes the oil a carrier of static. As per Noria Corporation, 2021, if an oil's

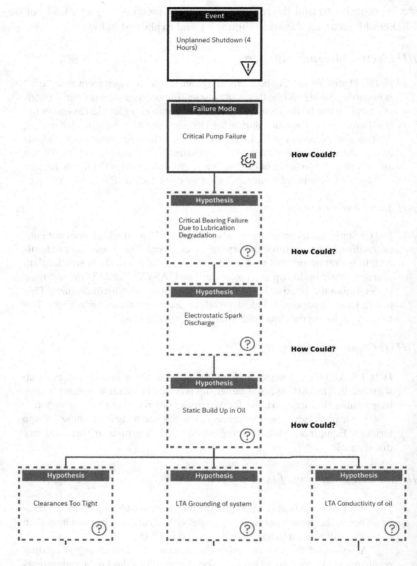

FIGURE 7.1 Lubrication degradation case – ESD (1).

conductivity is more than 400 picosiemens per meter at 20°C, there is little risk of ESD.

The ASTM D2624 could be performed and the conductivity of the oil could be matched against the thresholds noted by the lab.

We will now explore the hypothesis of "clearances being too tight" as shown in Figure 7.2. There are three main hypotheses for the clearances being too tight:

1) LTA OEM design for the system
2) Flow rate increased above the recommended operating threshold
3) Incorrect viscosity of the lubricant causing additional friction at a molecular level

For the purposes of easy graphical reference, only the first two hypotheses will be followed below.

[H] (1) LTA OEM design for the system

[VTC] The OEM may have designed the system for a particular set of applications. Any application falling outside of those specified may have issues with the clearance tolerances.

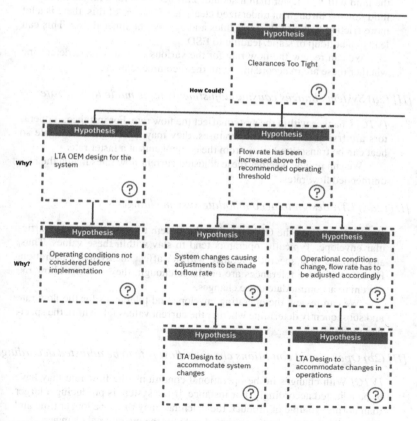

FIGURE 7.2 Lubrication degradation case – ESD (2).

We can review the recommended OEM applications and verify if these match the current applications. The OEM can also be contacted if any clarifications are needed.

[H] (1a) Operating conditions not considered before implementation

[VTC] OEMs can design systems for specific applications or conditions; however, when these systems enter the real world, those conditions can change. As such, particular conditions (such as temperature, type of fluid, or application) can all be impacted by the clearances. For instance, if the ambient conditions are colder than expected, the metal may contract, causing the clearances to become closer than before. As such, the potential for static buildup increases, which can lead to ESD.

[H] (2) Flow rate has been increased above the recommended operating threshold

[VTC] If the flow rate is increased above the recommended threshold, then the fluid will flow faster than it should. This accounts for a larger volume of fluid flowing through an undersized clearance. Because of this, there is a lot more friction between the molecules and against the inner lining. This can lead to a buildup of static, leading to ESD.

We can review the flow rate for the various clearances and determine whether these are in accordance with the recommendations.

[H] (2a) System changes causing adjustments to be made to flow rate

[VTC] Changes within the system affect the flow rate. For instance, if operators are trying to reduce temperatures, they may increase the flow rate so heat can be transmitted away from the components at a faster rate.

We can review whether these adjusted current flow rates match the recommended flow rates.

[H] (2a.i) LTA design to accommodate system changes

[VTC] Sometimes the OEMs design the system to operate within a particular envelope. Typically, operators tend to stay within these values. Thus, any changes of a specific tolerance will not affect the system's operation. However, if these tolerances are not large enough, they will not allow the system to accommodate these changes.

We can review the operating envelope and tolerances for the flow rate and subsequently determine whether the current values fall within the specified range.

[H] (2b) Operational conditions change – flow rate to be adjusted accordingly

[VTC] With changes in the operational conditions, the flow rate may have to be adjusted accordingly. For instance, if the system is producing a larger than normal volume of product, the machines may run for a longer time and require different flow rates to compensate for the operational change.

We can check the current flow rates, compare those to the recommended, and interview operators to find out whether any changes were made recently.

[H] (2b.i) LTA design to accommodate changes in operations

[VTC] The OEM may not have designed the system to accommodate these changes or the operators have been pushing the system outside of its normal operating conditions. Sometimes the OEMs design certain clearances to be expanded to accommodate these changes. As such, the clearance is no longer too small to accommodate the changes.

We can verify if changes to the clearances need to be adapted should the flow rates exceed the values currently being used in the system.

Let's take a look at the last hypothesis under the clearances being too tight. This hypothesis looks at whether the incorrect viscosity of the lubricant causes additional friction at a molecular level. When investigating this hypothesis, we can develop three more hypotheses, namely

1) OEM recommendations not followed
2) Unavailability of specified viscosity of the lubricant
3) LTA procedure for selecting the correct viscosity of the lubricant

Let's take a closer look at these in Figure 7.3 and below!

[H] Incorrect viscosity of the lubricant causing additional friction at the molecular level

[VTC] In ESD, static builds up at a molecular level. If the viscosity of the fluid is higher, then the number of molecules will increase. However, if the clearance remains the same then we would be forcing more molecules to fit in a smaller space. This can lead to additional friction among the molecules causing static buildup.

We can verify the current viscosity against the recommended viscosity to ensure that the incorrect viscosity is not being used.

[H] (1) OEM recommendations not followed

[VTC] If the OEM recommendations were not followed, then the incorrect viscosity could have been used. This would lead to the buildup of static, which could eventually lead to ESD.

We can verify by checking the OEM's recommendations and compare these to what is currently being used.

[H] (1a) OEM recommendations were not documented

[VTC] One of the reasons for the OEM recommendations are not being followed is that they were not documented and readily accessible to the users.

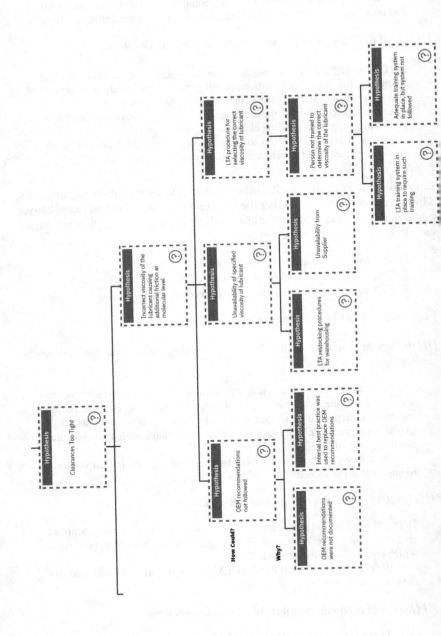

Most OEMs recommend the viscosity of the lubricant based on the operating conditions (such as temperature, speed, or load in some cases).

We can review the OEM manuals to determine whether these were documented or not.

[H] (1b) Internal best practice was used to replace OEM recommendations

[VTC] Internal best practices may have been used to replace OEM guidelines. This may be in an attempt to standardize or consolidate lubricants within the plant. Other users may have switched to a heavier viscosity and did not experience any failures hence this viscosity was used.

We can interview operators to find out the reason for the change in viscosity.

[H] (2) Unavailability of specified viscosity of the lubricant

[VTC] A heavier viscosity could have been used if the required viscosity were not available. Most suppliers suggest using a heavier viscosity instead of a lighter viscosity depending on the application. Typically, a heavier viscosity may be substituted in areas where wear is a higher concern than the speed or load-carrying capabilities of the oil.

We can verify the availability of the lubricant from our warehousing department.

[H] (2a) LTA restocking procedures for warehousing

[VTC] If the stock was depleted from the warehouse before they could restock then the procedure for restocking should be investigated. This may happen if there is an unexpected request for the lubricant or the usage of the lubricant was not forecasted and placed in the site's order.

The restocking procedure for the warehouse can be investigated as well as the trends of usage for the lubricant in question.

[H] (2b) Unavailability from supplier

[VTC] If the supplier does not have the required grade, a heavier viscosity may be used until the requested grade is made available. Suppliers can have periods of unavailability due to logistical challenges or issues at the blending plant and not being able to supply the requested volumes of lubricants.

The supplier can be contacted to verify if there is a shortage of the required grade.

[H] (3) LTA procedure for selecting the correct viscosity of lubricant

[VTC] Another way in which a heavier viscosity could have been used is if the wrong viscosity was selected. This can occur if the procedure for selecting the lubricant did not consider all the required factors (such as flow rate or application).

A review of the procedure for selecting the viscosity of the lubricant can be conducted.

[H] (3a) Person not trained to determine the correct viscosity of the lubricant

[VTC] If the person responsible for selecting the viscosity of the lubricant for the system is not trained, then they can select the wrong viscosity. The required viscosity is usually outlined in the OEM manual but there are instances where a viscosity–temperature chart is used. In these instances, based on the temperatures, the viscosity is selected.

We can review the training records of the person responsible for selecting the viscosity of the lubricant.

[H] (3a.i) LTA training system in place to require such training

[VTC] Some plants may not be aware of the training required for someone to be able to select the correct viscosity of lubricants. As such, the person responsible cannot make the request for training simply because they do not know if it exists.

We can review training records and interview the HR department on whether training sessions of this nature happened in the past.

[H] (3a.ii) Adequate training system in place, but system not followed

[VTC] Another reason for the person not being trained could be the system is not being followed. This can be because of budget constraints (lack of funds for training), shortage of personnel (so someone cannot leave for training), or the manager not approving the training.

We can review HR's training system and determine whether it is being followed.

Another hypothesis regarding the buildup of static is the absence of an adequate grounding system, as shown in Figure 7.4. The two main hypotheses for this LTA grounding system are:

1) Grounding doesn't exist
2) Grounding did not meet the OEM's requirements

Let's explore these in more detail below.

[H] LTA grounding system

[VTC] As explained earlier, having a LTA grounding system translates that there is no path through which the built up static in the oil can safely discharge. As such, the static will discharge on the filters causing the burnt membrane patches. We can determine the adequacy of the grounding system by consulting Subject Matter Experts (SMEs) in this field.

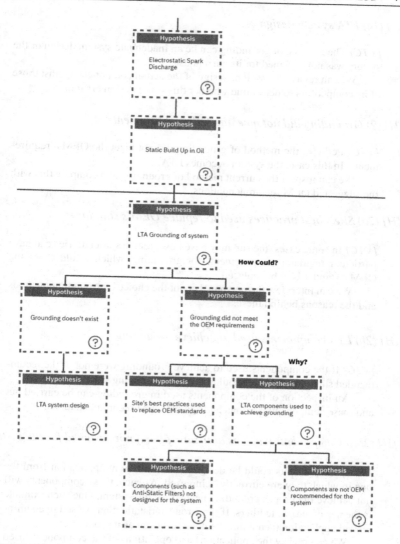

FIGURE 7.4 Lubrication degradation case – ESD (4).

[H] (1) Grounding doesn't exist

[VTC] If grounding does not exist, then this will be inadequate. Every system is supposed to have a path to the ground, through which any buildup of static can be discharged. It is a safety feature for many components. We can inspect the component and determine whether any grounding exists.

[H] (1a) LTA system design

[VTC] The absence of grounding can be an inadequate system design or the system was not designed for its current use.

We can review the system design of the actual component against those of its competitors to determine whether this was a design error or not.

[H] (2) Grounding did not meet the OEM requirements

[VTC] Perhaps, the method of grounding did not meet the OEMs' requirements. In this case, the system becomes LTA.

We can inspect the current method of grounding and compare this with the suggested OEM recommendations.

[H] (2a) Site's best practices used to replace OEM's standards

[VTC] In some cases, the site may have best practices or countries can have particular regulatory requirements for grounding, which would cause the OEM's standards to be replaced.

We can interview the personnel about the choice of grounding methods and the reasons behind these.

[H] (2b) LTA components used to achieve grounding

[VTC] If the components used to achieve grounding were not of the recommended standard, then this could make the grounding system to be LTA.

An inspection of the components used in grounding can be carried out and these components are matched against OEM recommendations.

[H] (2b.i) Components (such as antistatic filters) not designed for the system

[VTC] Components could be designed for a different application from the one in which they are currently being used. As such, these components will not achieve the proper grounding needed for the system. One such example is the use of antistatic filters. If the wrong antistatic filter is used in an unintended application then it may be not as efficient as it should be.

We can review the applications and operations of the components used in the grounding system to ensure these are designed for the system.

[H] (2b.ii) Components are not OEM-recommended for the system.

[VTC] Replacement components can be used instead of OEM-recommended to reduce cost. Therefore, these components are not recommended for the system and will not perform their intended functions.

We can interview the purchasing department to determine the method of selecting the replacement components and the reasons involved.

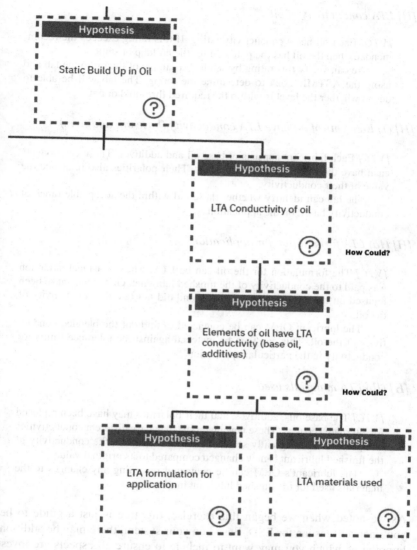

FIGURE 7.5 Lubrication degradation case – ESD (4).

We will finally explore the last hypothesis of the LTA conductivity of oil, as shown in Figure 7.5. This hypothesis leads to the elements of the oil having LTA conductivity which, in turn, can lead to the buildup of static in the oil. Let's explore this in more detail below.

[H] LTA conductivity of oil

[VTC] Each oil has a conductivity rating. If this rating exceeds the recommended, then the oil has the potential for the buildup of static.

We can verify this rating by sending a sample of the oil to the lab and using the ASTMD 2624 to determine the rating. The lab will be able to detect whether the level is within the required threshold or not.

[H] (1) Elements of oil have LTA conductivity (base oil, additives)

[VTC] Each lubricant consists of base oil and additives. These components each have their own conductivity levels. Their polarities also influence the value of their conductivity.

The lab can identify whether these fall within the acceptable range of conductivity based on the application.

[H] (1a) LTA formulation for application

[VTC] The formulation for the oil can be LTA, where not much attention was paid to the conductivity of the finished lubricant. OEMs may have been focused on the performance of the oil and did not rate the conductivity of the oil.

The lubricant OEM can be contacted to find out the blended conductivity of the oil. This can then be matched against the tolerance limits for conductivity in the particular application.

[H] (1b) LTA materials used

[VTC] The base oils, additives, and their polarities may have been replaced for alternate versions. These new materials may have different conductivities compared to the originally selected elements. As such, the conductivity of the finished lubricant is now changed compared to its original value.

The lubricant's OEM can be contacted regarding any changes to the materials used for the finished lubricant in question.

As we noted when we began this exercise, this tree is just a guide to help you uncover reasons for ESD. Based on your system, there may be additional hypotheses which you may want to include to ensure all aspects are investigated thoroughly.

Building a Tree for the LDM: Additive Depletion

8

WHAT IS ADDITIVE DEPLETION?

As mentioned in the last chapter, additives are sacrificial in nature. Very often, they get depleted while protecting the base oil. However, there are times when the additive can react with other components in the oil and come out of the solution to produce a deposit. Two main types of deposits exist, namely, organic or inorganic.

Organic deposits usually react to form a primary antioxidant species. These are typically rust and oxidation additives. On the other hand, inorganic deposits do not react with anything in the oil when they drop out. One example is zinc dialkyl dithiophosphate (ZDDP), which is used to reduce wear in lubricants.

WHAT LAB TESTS CAN BE PERFORMED?

When performing lab tests for the additive depletion degradation, the user will be looking for both the presence and absence of particular elements. The Fourier Transform Infrared (FTIR) test can help us establish the presence or absence of additives while the color test can give us an indication of which additive may have dropped out. Additives sometimes have distinctive colors, which may be seen in the color test. These are usually known to the lab and can be identified accordingly.

The quantitative spectrophotometric analysis (QSA) has been designed to identify insoluble molecules. This can assist users in identifying an

DOI: 10.1201/9781003252030-8

insoluble material which may be related to a particular additive. Additionally the Remaining Useful Life Evaluation Routine (RULER) test and Rotating Pressure Vessel Oxidation Test (RPVOT) can measure the loss of antioxidant additives in the oils (as previously described).

WHAT CAN BE DONE TO PREVENT ADDITIVE DEPLETION?

Normally, additives do not drop out of the oil. They only become displaced if triggered by environmental or operational conditions. However, if the finished lubricant was not blended according to the required specifications, then the additive can become easily displaced. Therefore, one way of avoiding additive depletion is to ensure that the finished lubricant meets the required specifications. Another method of reducing the potential for additive depletion is to guarantee that the lubricant does not come in contact with external products. This would avoid contamination and is covered in the next section.

APPLYING RCA TO AN ADDITIVE DEPLETION FAILURE

Additive depletion is not a very dominant lubrication degradation mechanism, however, it still occurs. As such, we should be able to identify it and determine the root causes for its occurrence. Let's discover some of the potential root causes for this failure mechanism.

As we've seen in the previous chapters, we begin each investigation with the reason we care. In this case, an unplanned shutdown of four hours is that top box as shown in Figure 8.1. This led to a critical pump failure which resulted from a critical bearing failure due to lubrication degradation. We are exploring additive depletion as a hypothesis in the example below.

[H] Additive depletion

> *[VTC]* Additive depletion is one of the modes of lubrication degradation. Confirmation of this mode is dependent on the deposits formed or the absence of particular additives. Deposits can be organic or inorganic in nature, and once tested by the lab, the presence of these additives in the deposits can be determined. On the other hand, if we test the oil, we will be looking for the absence of additives or their depletion.

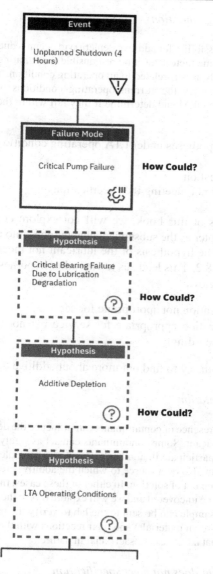

FIGURE 8.1 Lubrication degradation case – additive depletion (1).

[H] LTA operating conditions

[VTC] One possibility for additive depletion to occur is due to LTA operating conditions. Some factors must be responsible for agitating the additives out of oil. Typically, this is related to an operating condition.

We can verify the current operating conditions with those recommended by the OEM and determine if they fall within the expected parameters or not.

The two main hypotheses under LTA operating conditions include:

1) Contamination
2) Lubricant not meeting the specifications

For the purposes of this book, we will not explore contamination in much detail in this chapter as the subsequent chapter goes into more detail. However, we will explore the hypothesis of the lubricant not meeting specifications, as shown in Figure 8.2. This leads us to two other hypotheses, which carry us on our journey as below:

2a) Specification not appropriate for service
2b) Specification appropriate for service but not received from supplier accordingly

Let's start our journey to find out more about additive depletion below.

[H] (1) Contamination

[VTC] The presence of contaminants can influence the depletion of additives from the lubricant. Some contaminants can act as catalysts which cause the additives to participate in another reaction and be depleted. Similarly, contaminants can also be a source to which the additives react causing them to deplete or drop out of solution. In either of these cases, the source of contamination must be uncovered and we will explore this in the subsequent chapter.

An oil sample can be sent to the lab to verify the presence of contaminants (any foreign materials) and their reactions with the additives. We must remember that additives are sacrificial in nature.

[H] (2) Lubricant does not meet specification

[VTC] Another reason for additive depletion can reside in the fact that the lubricant did not meet its required specification. This specification can be related to its quality or actual performance.

In the instance of quality, if not blended to the required specifications, additives can easily be depleted from the oil.

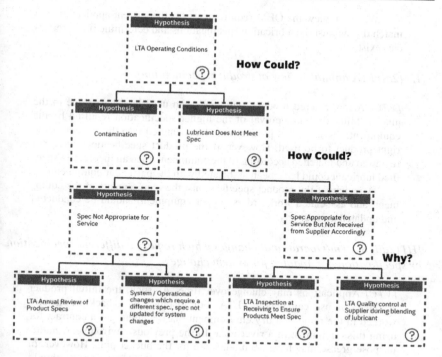

FIGURE 8.2 Lubrication degradation case – additive depletion (2).

In the instance of actual performance, the lubricant could be used in an application that requires a higher-performing lubricant. For example, hydraulic oil was used in an extreme environment where the presence of wear is a challenge. If the hydraulic oil was not blended with the adequate amount of ZDDP or not blended to its right specification, the ZDDP would easily be depleted from the oil.

In both of these instances, a certificate of analysis can be obtained from the OEM to determine whether this particular batch of lubricant was blended according to the required specification.

[H] (2a) Specification not appropriate for service

[VTC] As mentioned in the example above, the specification could be inappropriate for the service or application of the lubricant. If the specification does not meet the requirements of the environment in which the lubricant is working, then the additive can be depleted at a faster rate than expected. We must remember additives are sacrificial in nature. As such, their main purpose is to perform their function while being depleted.

We can review the OEM requirements for the current application and match this against the lubrication specifications and determine if discrepancies exist.

[H] (2ai) LTA annual review of product specifications

[VTC] A lower-tiered product may have been used in the past due to the unavailability or non-approval of the higher specification required by the equipment. As such, this product may have entered into the system as the right product to be used. However, if the product specifications were not reviewed and matched according to the equipment, then an incorrectly specified lubricant could have been associated with the particular equipment.

A review of the product specifications, the requirements of the equipment, and associated work orders (as per equipment) should be conducted and updated accordingly.

[H] (2aii) System/Operational changes which require a different specification or specification not updated for system changes

[VTC] Applications may change over time. As such, the initial product selected for use can become LTA. For instance, if a backhoe usually worked in the mine in a dry environment but the client got a contract for using this equipment in a river area. In the previous environment, multipurpose grease may have been adequate for the bucket pins. However, in this new environment, with the presence of water, a different grease should be used (such as something containing a calcium complex thickener) to protect the pins.

We can review the specifications of the lubricant based on its application and determine whether an update on the specifications was made or not.

[H] (2b) Specification appropriate for service but not received from supplier accordingly

[VTC] The product received could not meet the specifications as indicated by the supplier. Sometimes, suppliers may deliver an incorrectly specified product due to challenges experienced on their supply side. We can verify if the product received matches the specification required by reviewing the certificate of analysis for the particular batch of product.

[H] (2bi) LTA inspection at reception to ensure products meet specification

[VTC] At the customer's receiving site, there could have been LTA checks to ensure the product meets the required specification. Upon receipt of the product, customers should verify they are receiving the correct product as indicated by the labeling on the product. Sometimes, there may be a mistake at the supplier's warehouse and they send the wrong grade of product or an entirely different product. It is the duty of the warehousing staff at the customer's site to ensure they are receiving the correct product.

We can review the procedure for receipt of product from suppliers and determine whether there are tasks that ensure proper checks are done upon receipt.

[H] (2bii) LTA quality control at supplier during blending of lubricant

[VTC] At the supplier's site, there could have been a quality control issue (lubricant not blended at the right temperature so that additives settle out after some time). These can be verified by reviewing the certificate of analysis for the particular batch of lubricant. This is usually provided by the lubricant's supplier.

The aforementioned is to be used as a guide for understanding some root causes behind additive depletion. This can be expanded with more hypotheses depending on your system and observations in your investigation. As mentioned earlier, contamination is not covered in detail for this tree as we will be covering it in further detail in the next chapter.

Building a Tree for the LDM: Contamination

9

WHAT IS CONTAMINATION?

Contamination can be described as the ingress of any foreign material into a system. Typically, these foreign materials act as catalysts which can promote other types of degradation mechanisms. These foreign objects are normally classified into three main groups: metals, air, and water. Contamination can be a precursor to oxidation, microdieseling, thermal degradation, and even additive depletion.

WHAT LAB TESTS CAN BE PERFORMED?

The first sign of contamination can usually be seen visually. For example, the lubricant does not appear as its normal color (perhaps the lubricant appears green when it should be golden) or consistency (if there are items trapped in the lubricant, it can increase its viscosity). Generally, if the foreign items can be seen with the naked eye, it means that the contamination levels are very high!

The color test can be used as a guideline in establishing whether there are foreign items present. Additionally, there are tests for the presence of water, fuel, or coolant which return numeric values. In diesel engines, users usually see fuel dilution but there are limits which must be observed (depending on the OEM), where the ingress of fuel will harm the lubricant.

DOI: 10.1201/9781003252030-9

WHAT CAN BE DONE TO PREVENT CONTAMINATION?

Contamination involves the ingress of foreign materials into the system. These materials can enter the system from outside or inside. Hence, users need to ensure that there are no leaks or openings to allow these materials to enter the system. If the material is coming from inside the system, then the user needs to establish boundaries to reduce this ingress.

APPLYING RCA TO A CONTAMINATION-RELATED FAILURE

Contamination is one of the most interesting degradation mechanisms. Its presence can lead to many of the degradation mechanisms discussed in the previous chapters! Hence, being able to identify its root causes can potentially impact the initiation of many other degradation mechanisms. Let's explore the thought process behind finding these root causes!

As usual, we begin with our top box or the reason we care. In this case, we had an unplanned shutdown which occurred due to a critical pump failure, as shown in Figure 9.1. The critical pump failed since a critical bearing failure occurred due to lubrication degradation. In this tree, we will be investigating the hypothesis of contamination below.

[H] Contamination

[VTC] Contamination can only occur if there is an ingress of foreign materials into the lubricant. There are a few different tests which can be used to confirm the presence of contamination. The oil can be sent to the lab to verify and quantify the presence of water, fuel, or other elements. We must remember that contamination can also occur in the form of metals. For instance, if an oil sample is taken and the results show the presence of nickel but there is no nickel in the system then this is a foreign material which has entered the system. We can also detect the presence of foreign materials by visually inspecting the lubricant. There may be some elements which are not visible to the human eye and the lab can test for these.

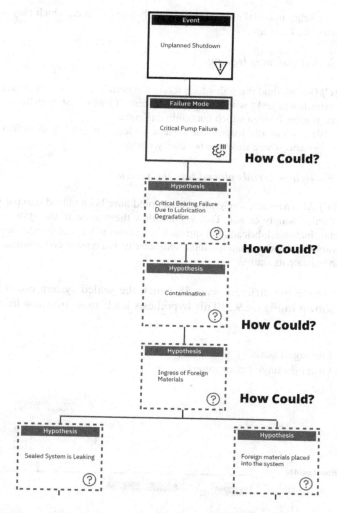

FIGURE 9.1 Lubrication degradation case – contamination (1).

[H] Ingress of foreign materials

[VTC] The main form of contamination is the ingress of foreign materials. These are materials which should not be in the system. Their presence can initiate reactions in the lubricant, which can lead to further degradation.

We can send a sample of the lubricant to the lab for confirmation on the type of foreign materials. We can also investigate the filters to determine if any foreign material entered the system before the filter, which may have affected the lubricant.

[H] (1) Sealed system is leaking

[VTC] One method through which foreign materials can enter the system is the existence of leaks within the sealed system. If leaks exist, then this could be the source through which contaminants enter.

We can visually inspect the system for leaks or use UV detection for those systems where leaks are not easily visible.

[H] (2) Foreign materials placed into the system

[VTC] Alternatively, foreign materials could have been placed into the system unknowingly or not. This would allow them to be in the system and contaminate the lubricant. Oil sampling at various points can help to identify possible sources of contaminants. Additionally, the type of contaminant can also indicate its source.

Let's examine the different ways in which the sealed system could be leaking, as shown in Figure 9.2. This hypothesis leads us to two new hypotheses, namely

1) Damaged seals
2) Other damaged components

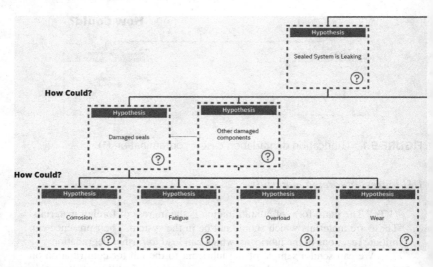

FIGURE 9.2   Lubrication degradation case – contamination (2).

Typically, seals are common in most systems and are the barriers protecting the lubricant. However, other components should be investigated for damages through which contaminants can enter. Both seals and components can be considered mechanical items. As such, damages to either of them will fall under the broad categories of corrosion, wear, fatigue, and overload, which we will explore in greater detail below.

[H] (1) Damaged seals

[VTC] Seals are essential in most lubrication systems. They are responsible for keeping the lubricant in place and ensuring entry of no contaminants. Hence, if they become damaged, contaminants can enter the lubricant. Seals can be visually inspected, and their integrity is verified to ensure there is no passage of contaminants into the lubricant.

[H] (2) Other damaged components

[VTC] As noted above, seals are not the only components in which the lubricant interacts. Any other component (this will vary depending on the system) that becomes damaged can allow contaminants to enter the lubricant. Thus, all components need to be inspected to ensure there are no damages. A visual inspection can be done to achieve this verification.

[H] (1a) Corrosion

[VTC] This failure pattern is often recognized by "pitting". Corrosion is actually an electrochemical phenomenon. There are many types of corrosion, so such failure mechanisms should be reviewed by trained professionals to determine which type of corrosion is at play.

[H] (1b) Fatigue

[VTC] Fatigue is a cyclic force on the material that happens gradually over time. This is often evidenced by the observation of "beach marks"-like patterns on a failed surface. As a crack propagates, eventually the remaining material cannot take the load and there will be evidence of an instantaneous fracture zone (IFZ).

[H] (1c) Overload

[VTC] Overload is an instantaneous overpowering of the material. A ductile material will deform while a brittle material will fracture. Materials will show a salt and pepper appearance when a brittle fracture takes place. Any material that experiences loading above its ultimate tensile strength (UTS) will fail. Ductile and brittle are the primary types of overload. Again, a trained eye should evaluate and diagnose which failure mechanisms are at play, before drilling deeper to determine the "Whys".

[H] (1d) Wear

[VTC] Wear is the uniform loss of material across a fractured surface. This can be caused by a number of conditions where two surfaces meet and rub resulting in a loss of metal. Fractured components should be properly preserved in a temperature-controlled staging area to prevent further damage. This is until a determination is made by a trained eye (i.e., metallurgist/materials engineer), who will inspect it and/or ship the failed parts to a certified metallurgical lab. This is true for fatigue, corrosion, and overload as well.

One of the methods in which foreign materials can be placed in the system is the input of wrong lubricants into the system, as shown in Figure 9.3. There are two hypotheses which can be applied here.

1) Incorrect lubricant selection
2) Unclean components – previously exposed to another lubricant

For the ease of graphical representation in this tree, we will explore the hypothesis of incorrect lubricant selection. For this hypothesis, there are three more hypotheses:

1a) Warehouse dispatched the wrong lubricant
1b) Operations/maintenance selected the wrong lubricant
1c) OEM supplied the wrong lubricant

We will now explore these below.

[H] Wrong lubricant entering the system

[VTC] If the wrong lubricant is placed in the system, then this is a foreign material as it should not be there. The wrong lubricant can lead to improper viscosity, leeching of additives, or promotion of reactions, which should not occur.

We can send an oil sample to the lab to confirm the identity of the lubricant in the system and then match this to the lubricant which should be in the system.

[H] (1) Incorrect lubricant selection

[VTC] The incorrect lubricant could have been selected causing the wrong lubricant to enter the system. We can verify this through interviews with the personnel responsible for selecting the lubricant.

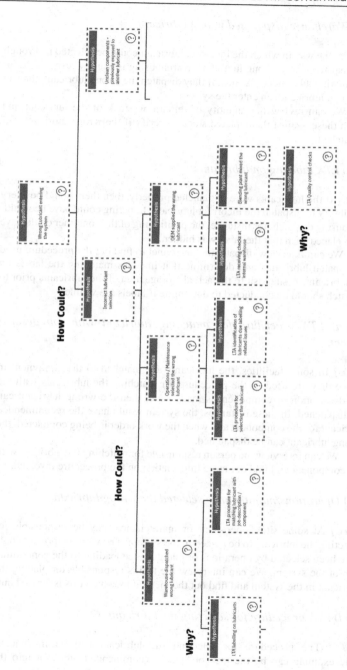

FIGURE 9.3 Lubrication degradation case – contamination (3).

[H] (1a) Warehouse dispatched wrong lubricant

[VTC] One way in which the incorrect lubricant could be selected is through the dispatch of the wrong lubricant. Warehousing is usually responsible for dispatching lubricants. As such, if they dispatch the wrong lubricant, then an incorrect lubricant can enter the system.

We can review the quantity of remaining stock of the lubricant and match these against the depleted stock to verify if the wrong lubricant was dispatched.

[H] (1a.i) LTA labeling on lubricants

[VTC] If the lubricants were not labeled properly, then there could be every possibility to dispatch the incorrect lubricant. Labeling could have been hidden during dispatch, for instance, the positioning of the container could have been placed such that the label was hidden.

We can interview dispatching personnel to find out the procedure used to dispatch lubricants and determine if it includes matching the labels on the lubricant. A site visit to the actual storage area of the lubricants prior to dispatch should be conducted to determine if labels are visible.

[H] (1a.ii) LTA procedure for matching lubricant with job description/component

[VTC] In some facilities, the lubricants are matched to the equipment in which they are used. If the procedure for matching the lubricants with the job description or component is inadequate, then the wrong lubricant can be dispatched. In these instances, the system would have the recommended lubricant for the component and when the work order is being completed, the wrong lubricant can be dispatched.

We can interview the person responsible for matching the lubricant with the component and find out more information on the procedure involved.

[H] (1b) Operations/maintenance selected the wrong lubricant

[VTC] At some sites, operations or maintenance may be responsible for selecting the lubricant to be placed in the system. The wrong lubricant could have been selected by operations or maintenance leading to the contamination of the system. We can interview the teams responsible for placing the lubricant in the system and find out the method by which this is carried out.

[H] (1b.i) LTA procedure for selecting the lubricant

[VTC] The procedure for selecting the lubricant can be inadequate. One example can be the task of "place recommended lubricant into the equipment". This task leaves the person with the choice of choosing any lubricant they believe to be recommended. Because of this, the wrong lubricant can enter the equipment.

We can review the procedure involved in selecting the lubricant and adjust this accordingly.

[H] (1b.ii) LTA identification of lubricants due to labeling-related issues

[VTC] Should the labeling of the lubricants be inadequate then the wrong lubricant can be selected. One example is the labeling of hydraulic oils as "Hydraulic oil" with no further description of the viscosity of the oil. This typically occurs with decanters left out in the field. If proper labeling doesn't exist, then the wrong lubricant can enter the system and cause contamination.

[H] (1c) OEM supplied the wrong lubricant

[VTC] Another way in which the wrong lubricant could enter the system is if the OEM supplied the wrong lubricant. Sometimes, this can occur, and the customer may not be aware of the wrong lubricant being supplied. As such, this can be placed into the system causing contamination.

We can review orders placed to the OEM and match these against the received product to determine whether there are any discrepancies.

[H] (1c.i) LTA receiving checks at internal warehouse

[VTC] If the OEM supplied the wrong lubricant, the warehouse may have been able to identify this if they had adequate checks upon receipt. Warehousing personnel responsible for receiving products should always ensure they receive the products which were ordered.

We can review the procedure for receipt of the product by the site to ensure adequate verifications are conducted upon receipt of the product.

[H] (1c.ii) Blending plant mixed the wrong lubricant

[VTC] The blending plant could have blended the incorrect product (blended a gear oil when it should have been hydraulic or mixed and incorrect ratio of elements) but the error was not detected.

The certificate of analysis should have been able to reflect any discrepancies. Additionally, an oil sample could be taken from a new drum/container of lubricant and sent to the lab to aid in identifying the type of product received.

[H] (1c.ii1) LTA quality control checks

[VTC] The certificate of analysis should have reflected any discrepancies. However, if this is not verified before leaving the OEM's facility, this could be the source of the LTA quality control checks.

An interview can be conducted with the OEM to discuss the quality control checks which are used in the facility.

We will now explore the other hypothesis of the wrong lubricant entering the system, as shown in Figure 9.4. An unclean component which was previously exposed to another lubricant can be a source of the wrong lubricant entering the system. There are two hypotheses which can be derived, namely

1) Components were used from another system with a different lubricant or
2) Installation/maintenance tools exposed to another lubricant.

Let's explore these hypotheses below!

[H] Wrong lubricant entering the system

[VTC] Foreign materials entering the system can be the wrong lubricant. As explained in the earlier table, the wrong lubricant is classified as a foreign material. We can verify the presence of the wrong lubricant by sending an oil sample to the lab.

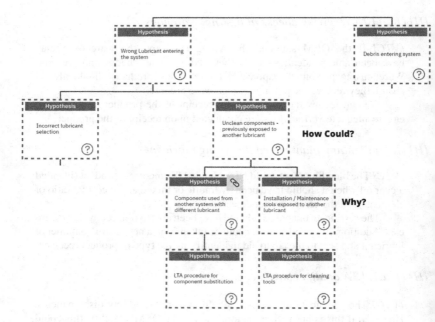

FIGURE 9.4 Lubrication degradation case – contamination (4).

[H] Unclean components – previously exposed to another lubricant

[VTC] One way of getting the wrong lubricant in the system is through the transfer of lubricant via components. If a component in the system was previously used in a different system and is transferred to another system without being cleaned, then there is a possibility of transferring the wrong lubricant to the new system.

This can be the case for pipes or components which may have their own reservoirs. If transferred from a system which used an ISO 46 hydraulic oil to a system using an ISO 32 hydraulic oil, then the heavier oil (depending on the quantity) can impact the performance of the system. There can be a viscosity change and the system's oil becomes contaminated.

We can verify whether there were any replacements of components in the system from interviewing the personnel working on the system.

[H] (1) Components used from another system with different lubricant

[VTC] The components being replaced may have been removed from another system (perhaps a system which was no longer in use). Perhaps, the need arose for this particular component in the actual system and the component was replaced without thinking of cleaning it before placing it in the system.

Interviews can be conducted, and work orders reviewed to determine whether any components were replaced.

[H] (1a) LTA procedure for component substitution

[VTC] If a procedure for component substitution does not exist or it exists but LTA, then an unclean component can be placed in the system. Sometimes, when personnel is rushing to get production restarted, decisions can be made, which may seem harmless at the time. This can be compounded with the absence of clear instructions regarding the replacement of components.

A review can be conducted of the existing procedure for substitution of parts or, in its absence, an interview with the supervisor regarding the instructions given in such circumstances.

[H] (2) Installation/maintenance tools exposed to another lubricant

[VTC] Another way of transferring the wrong lubricant could be through the maintenance tools. If these were not cleaned after use in another environment, then they can transfer the wrong lubricant to another system. Let's think of this as a surgeon preparing for surgery. After each operation, all tools are sanitized in preparation for the next operation. We wouldn't want tools which were not cleaned being used on us! It should be the same for machines.

We can interview the personnel involved in installation or maintenance to find out their practices of cleaning tools or not cleaning tools after use.

[H] (2a) LTA procedure for cleaning tools

[VTC] If there is a LTA procedure for cleaning tools or a very vague instruction such as "Put away tools when finished", then there is no mention of cleaning the tools. We should review the procedures to determine whether the procedure is adequate.

Another method through which foreign materials can enter the system is through debris entering the system, as shown in Figure 9.5. If debris enters the system, there are two main hypotheses which follow

1) Damaged filter membrane
   a. LTA filter
   b. Alert system malfunctioning, no alert received
   c. Wrong installation of filter
2) Unprotected opening to the system

For ease of graphical reference, we will explore the hypothesis of a damaged filter membrane and LTA filter below.

[H] Debris entering system

[VTC] Foreign materials can also enter the system in the form of debris or solid particles.

We can verify this presence either by a visual inspection (where we can see the debris visible to the naked eye) or through an ISO particle count test with the lab. The latter can quantify particles which are not visible to the naked eye.

[H] Damaged filter membrane

[VTC] One of the main purposes of a filter is to remove unwanted particles. However, if the filter membrane is damaged then the filter will no longer perform its function. As such, debris can easily enter the system.

We can verify the condition of the membrane by performing a visual inspection of the filter membrane or taking oil samples before and after the filter and then comparing the results. This helps us to understand the efficiency of the filter in its system.

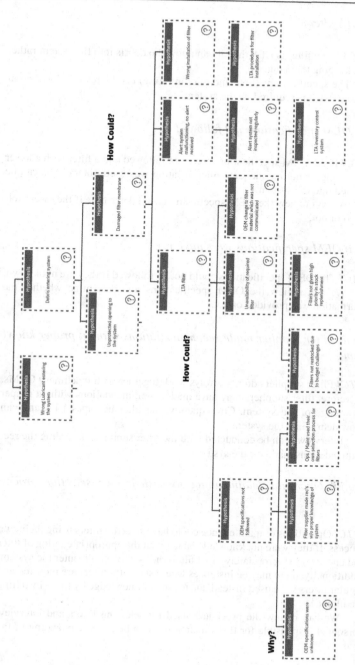

FIGURE 9.5 Lubrication degradation case – contamination (5).

[H] (1) LTA filter

> *[VTC]* If the filter is LTA then it can allow the debris into the system rather than keeping them out.
>
> The specifications (beta and micron rating) can be reviewed and compared to what is required by the system.

[H] (1a) OEM specifications not followed

> *[VTC]* If the OEM specifications were not followed and a filter with a lower beta rating or not an adequate micron rating was used, then debris can be allowed into the system.
>
> We can review the OEM specifications and determine if they were followed or not.

[H] (1a.i) OEM specifications were unknown

> *[VTC]* The OEM specifications could not be followed if they are not known.
>
> We can review the OEM specifications and determine whether the information was provided or not.

[H] (1a.ii) Filter supplier made recommendations without proper knowledge of system

> *[VTC]* Filter suppliers do not always work together with machinery OEMs. Therefore, filter suppliers may have made recommendations without proper knowledge of the system. Consequently, the filter becomes LTA and can allow debris into the system.
>
> Interviews can be conducted with the filter suppliers regarding the recommendations made for the system.

[H] (1a.iii) Operations/maintenance used their own selection process for filters

> *[VTC]* Operations or maintenance could have selected filters using their own process. If they were not knowledgeable about the appropriate rating of filter and chose an inappropriately rated filter, then debris could enter the system. Additionally, there may be instances where specialty filters are required and regular filters were used instead due to lack of knowledge by those installing the filters.
>
> We can review the procedure used for selecting filters and interview personnel responsible for this function to get a better understanding of the process.

[H] (1b) Unavailability of required filter

[VTC] The required filter could be unavailable and hence a replacement filter is used, which led to the filter being LTA and debris entering the system.

We can investigate whether the required filters were available at the time of replacement.

[H] (1b.i) Filters not restocked due to budget challenges

[VTC] With budget challenges, the finance division may look at avenues for reducing expenditure. One way of achieving this can be prolonging the stock replenishment cycle. In this way, filters may not be available when required due to the decisions made by the finance team.

We can interview the finance team and find out if there were any budget cuts which may have affected the availability of the filters.

[H] (1b.ii) Filters not given high priority in stock replenishment

[VTC] Filters are sometimes viewed as nonessential items and may not be given a high priority in stock replenishment. As such, they could not be available when requested and another substitute filter is used.

We can review the priority listing for ordering items and determine if filters can be given a higher priority or a healthier stock is kept.

[H] (1b.iii) LTA inventory control system

[VTC] Should there be a LTA inventory control system whereby an accurate account of the quantity of filters in stock is not kept, then filters can be unavailable when requested.

We can review the inventory control system to determine if this is regularly updated.

[H] (1c) OEM's change of filter material was not communicated

[VTC] OEMs can make changes to the material being used in the filters and not inform customers. As such, this change in the material may not be compatible with the system flow and may allow debris to enter.

The OEM can be contacted regarding any changes in the material used for the filters.

[H] (2) Alert system malfunctioning, no alert received

[VTC] A damaged filter membrane usually triggers an alarm as the differential pressures in the system would change. However, if the alert system is malfunctioning and no alert is received then debris can be entering the system. In some cases, the filter can go into bypass, allowing debris and other contaminants to enter the system.

We can inspect the alarm system to ensure it is working.

[H] (2a) Alert system not inspected regularly

[VTC] Alarm systems are not used every day. As such, there can be instances where these alarm systems are not inspected on a regular basis to ensure they are working.

We can review the alarm inspection process and frequency to ensure the levels are adequate.

[H] (3) Wrong installation of filter

[VTC] The filter could have been installed incorrectly. There have been instances where filters were installed upside down due to lack of instructions from the supplier.

An interview can be conducted with personnel responsible for installing filters to find out the routine for installation. An inspection of the filters used can also be conducted to ensure these are easy to install with marginal room for error.

[H] (3a) LTA procedure for filter installation

[VTC] The procedure for installing the filter could be LTA. Most work order sheets simply state, "Install filter". However, detailed instructions of which side should be placed on the respective component and guidelines on the torque required to tighten the filters should be included.

A review of the procedure for the installation of filters should be conducted to determine if it is adequate.

Debris can also enter the system through an unprotected opening, as seen in Figure 9.6. We will explore the different ways in which this can occur, namely

1) Hatch covers left open
   a. Technicians did not close the covers after performing the task
   b. Previously damaged hinges or clasps prevented proper locking
2) Protective coverings of openings were damaged
   a. LTA inspection for protective coverings of openings

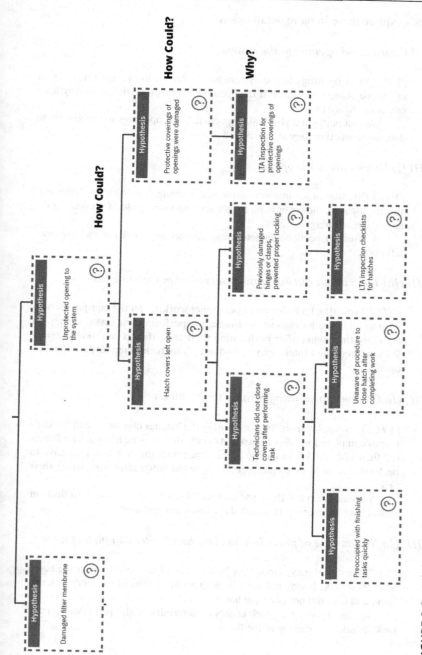

FIGURE 9.6 Lubrication degradation case – contamination (6).

Let's explore these in more detail below!

[H] Unprotected opening to the system

[VTC] Most openings to the system are protected to prevent the entry of unwanted objects. However, if the opening has no protection, then anything can enter the system.

We can perform a physical inspection of all openings to the system to determine whether they are protected or not.

[H] (1) Hatch covers left open

[VTC] One way of having an unprotected opening is to leave the cover off! In this instance, the cover is no longer serving its purpose of preventing the admission of foreign objects.

A visual inspection can determine whether any hatch covers have been left open.

[H] (1a) Technicians did not close covers after performing task

[VTC] Typically, technicians or personnel working on the equipment may have to open the hatches to get inside and perform their tasks. If they do not close the covers after performing their task, then it can remain open. We could conduct interviews to find out if any technician left the hatches open.

[H] (1a.i) Preoccupied with finishing tasks quickly

[VTC] Personnel responsible for opening the hatches may be on a tight deadline to complete a number of tasks. As such, they may be hurried into finishing their task quickly or can be preoccupied with the next task they have to perform. As such, they can forget to close the hatch after completing their task.

We can interview the personnel working on the equipment to find out more about the manner in which these tasks are performed.

[H] (1a.ii) Unaware of procedure to close hatch after completing work

[VTC] Personnel may also simply be unaware of the need to close the hatch after their task. It may not have been explicitly stated in the work order as such, and they did not close the hatch.

We can review the work orders to determine if the procedure for the task includes the closing of the hatch.

[H] (1b) Previously damaged hinges or clasps prevented proper locking

[VTC] The technicians could have closed the hatch but if the clasps are damaged and prevent proper locking, then they can remain open.

We can perform an inspection regarding the integrity of the hinges or clasps for the openings.

[H] (1b.i) LTA inspection checklists for hatches

[VTC] Upon inspection of the hatch, it should be determined whether the hinges work properly or not. This is not a simple visual inspection. However, if the inspections for the hatches did not include inspecting the hinges, then it can become LTA.

We can review the inspection checklists for hatches.

[H] (2) Protective coverings of openings were damaged

[VTC] Protective coverings cover the entire surface of the hatch. If these are damaged in any way, it can allow debris to enter the system. A visual inspection can reveal the integrity of these coverings.

[H] (2a) LTA inspection for protective coverings of openings

[VTC] If the inspection of the protective coverings of the opening is LTA, then this breach will be unnoticed. As such, debris will enter the system and the system can become contaminated. Some inspections for protective coverings can be very vague such as "Ensure the protective covering is in place". This does not speak to the integrity of the covering.

We can review the inspection procedure for the protective coverings to ensure it verifies their integrity.

In this example, we explored the various root causes of contamination as a lubrication degradation mechanism. These can be adapted to your system and become more specific. For instance, you can reference particular openings or processes which may cause contamination and insert these into the tree. This way, the tree becomes a template for your site and anyone using the tree will be familiar with the processes or assets which you inserted. The aim of this tree is to provide some guidance and additional nodes can be inserted as per your system.

Bringing It All Together: Fan Motor Failure RCA

10

BRINGING IT ALL TOGETHER

We have now described in detail each of the six primary mechanisms for lubrication degradation along with the summary of RCA logic tree templates related to its causes and contributing factors. Now we want to apply these "pieces of the puzzle" to a realistic case from the field.

CASE BACKGROUND

Statement of the Failure Event: On June 15, 20XX at 4:30 PM, production at the ABC widget factory had to be reduced when the quenching section could not meet the quality level of cooling needed. The cooling water temperature had increased due to the failure of the #2 cooling tower cell fan motor. The production department was required to slow the line down to meet the quality specification of the quenching section. It took the maintenance team five hours to replace the motor and return the #2 cooling tower cell back into service. With the cooling capacity of the tower fully restored, production was able to return to the normal level.

Consequences of the Failure Event: The impact of the reduced production level was USD 200,000 loss in revenue opportunity. The repairs cost another USD 44,000 for maintenance labor, crane rental, and contracted riggers. The motor was sent to a repair shop for rebuilding and testing, where the bearings

DOI: 10.1201/9781003252030-10

were retained for future examination. The cost for the repairs to the motor was USD 6,000.

DATA/EVIDENCE COLLECTION

As we break down this case study into its manageable components, we will be maintaining a basic verification log (VL) in a table format. This, at a minimum, will identify our verification methods and their resulting outcomes. It is highly recommended for the RCAs of any significant level of consequence to strictly maintain this verification log as it will be what makes your logic tree stand tall with confidence!

Typically the verification log takes the form of a table with the following headings:

- Hypotheses [H]
- Verification Technique [VT]
- Verification Outcome [VO]
- Date Completed (excluded in this example)
- Responsibility (excluded in this example)

The associated VL for the nodes identified in Figure 10.1 is shown below.

[H] Armature/winding failure

> *[VT]* Forensic review
> *[VO]* Not true

[H] Bearing failure

> *[VT]* Forensic review
> *[VO]* True

[H] Housing/casing failure

> *[VT]* Forensic review
> *[VO]* Not true

[H] Corrosion

> *[VT]* Forensic review
> *[VO]* Not true

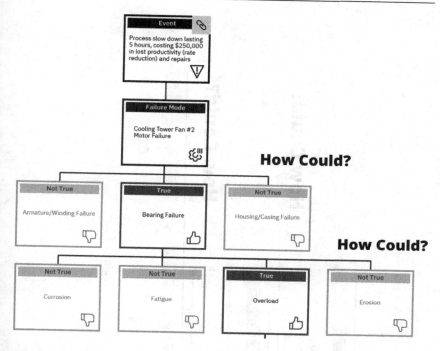

FIGURE 10.1   Levels 1–4.

[H] Fatigue

>   *[VT]* Forensic review
>   *[VO]* Not true

[H] Overload

>   *[VT]* Forensic review
>   *[VO]* True

[H] Erosion

>   *[VT]* Forensic review
>   *[VO]* Not true

Now let's continue drilling down this logic tree by asking the "How Could" questions until we get to the human root(s), or the decision points.

Below shows the associated VL for the nodes identified in Figure 10.2.

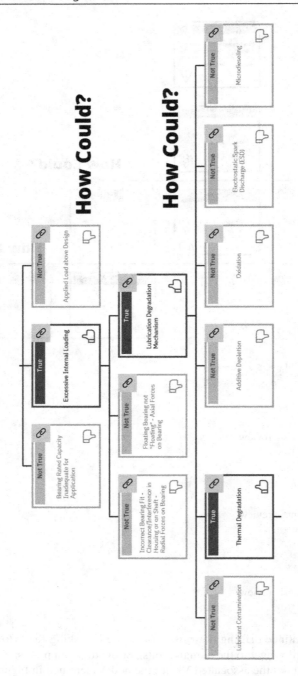

FIGURE 10.2   Levels 5–7.

[H] Bearing rated capacity inadequate for application

> *[VT]* Review of bearing specs
> *[VO]* Not true

[H] Excessive internal loading

> *[VT]* Inspect bearing for signs of damage to the race surfaces from internal loading stresses.
> *[VO]* True

[H] Applied load above design

> *[VT]* Review of design documents
> *[VO]* Not true

[H] Incorrect bearing fit – clearance/interference in housing or on shaft-radial forces on bearing

> *[VT]* Measure shaft diameter, housing bore diameter, bearing inner and outer diameters. Calculate actual fits and compare them to OEM specifications or SKF bearing design parameters.
> *[VO]* Not true

[H] Floating bearing not "floating" – axial forces on bearing

> *[VT]* Inspect bearings upon disassembly of unit to confirm bearing locator shim is only on one shaft bearing. Other bearings must be free to move in the axial direction.
> *[VO]* Not true

[H] Lubrication degradation mechanism

> *[VT]* Forensic analysis of failed bearing and lab analysis of lubricant
> *[VO]* True

[H] Lubricant contamination

> *[VT]* Lubricant sampling and analysis
> *[VO]* Not true

[H] Thermal degradation

> *[VT]* Lubricant sampling and analysis
> *[VO]* True

[H] Additive depletion

> *[VT]* Lubricant sampling and analysis
> *[VO]* Not true

[H] Oxidation

> *[VT]* Lubricant sampling and analysis
> *[VO]* Not true

[H] Electrostatic discharge

> *[VT]* Lubricant sampling and analysis
> *[VO]* Not true

[H] Microdieseling

> *[VT]* Lubricant sampling and analysis
> *[VO]* Not true

[H] Too much lubricant

> *[VT]* Look for grease being pushed out of the seal.
> *[VO]* Not true

[H] Too little lubricant (physical root)

> *[VT]* Review vibrational analysis and look for an increase in temperature.
> *[VO]* True

[H] Lubricant contamination

> *[VT]* Inspection of remaining and stored lubricants.
> *[VO]* Not true

We will continue drilling down levels 8–10 with our "How Could" questioning using our evidence to determine which legs we continue to follow.

Below shows the associated VL for the nodes identified in Figure 10.3.

[H] Improper lubrication replenishment

> *[VT]* Review for evidence found of starvation of the bearing through vibrational analysis and temperature trends.
> *[VO]* True

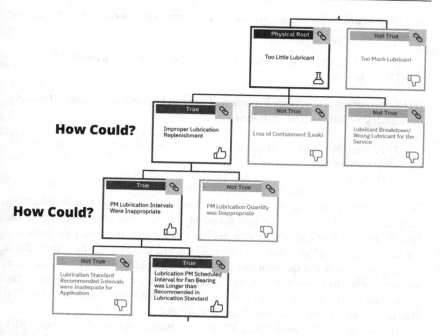

FIGURE 10.3 Levels 8–10.

[H] Loss of containment (leak)

> *[VT]* Review for evidence of leaks in the bearing.
> *[VO]* Not true

[H] Lubricant breakdown/wrong lubricant for the service

> *[VT]* Determine if the lubricant broke down. Cross-examine the lubricant used with the specified lubricant from the OEM to see if there is a match.
> *[VO]* Not true

[H] Preventive Maintenance (PM) lubrication intervals were inappropriate

> *[VT]* Compare the lubrication periods in the PM vs the specified lubrication intervals as per the OEM.
> *[VO]* True

[H] PM lubrication quantity was inappropriate

> *[VT]* Verify the quantity of the lubricant was allocated to the PM vs the quantity specified in the manuals for the conditions that exist (load, speed, and temperature).
> *[VO]* Not true

[H] Lubrication standard recommended intervals were inadequate for application

> *[VT]* Did the lubrication standards meet the OEM specified intervals?
> *[VO]* Not true

[H] Lubrication PM scheduled interval for fan bearing was longer than recommended in lubrication standard

> *[VT]* Were lubrication PM schedules extended based on standardizing for other components?
> *[VO]* True

Notice at this point that we start to see some labels on the nodes indicating they are a bit different. The key takeaway here is the identification of the human root in this case. This is a decision point that triggered the previous cause-and-effect relationships to progress toward the undesirable event. So, it is here we switch the questioning from "How Could" to "Why". Remember, we don't want to know "How Could" someone have made the decision at the time because the possibilities could be infinite. We want to know "Why" on that day, at that time, it seemed like the right decision.

Below shows the associated VL for the nodes identified in Figure 10.4. Note that in this table, there are corrective actions toward the end. These are our proposed countermeasures.

[H] Supervisor implemented a change in PM interval not following the lubrication standard (human root)

> *[VT]* Obtain PM histories and review for interval histories. Interview the supervisor around the reasoning for changing the PM interval unilaterally. Why did they feel it was the appropriate decision?
> *[VO]* True

[H] Production pressures forced a trade-off decision to be made (systemic root)

> *[VT]* Interview the supervisor. Have a discussion about how the day started and led up to a conversation about that decision, at that time. What triggered the diversion from the Standard Operating Procedures (SOP)?
> *[VO]* True

[H] LTA management oversight (systemic root)

> *[VT]* Interview the supervisor. During the discussion, ask about the decision that day and the reasoning. See if this was a typical practice that was condoned in the past.
> *[VO]* True

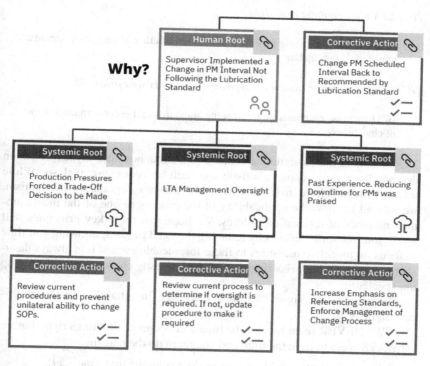

FIGURE 10.4 Levels 11–13.

[H] Past experience. Reducing downtime for PMs was praised (systemic root)

> *[VT]* Interview the supervisor and manager. Determine if not following the SOP was acceptable in the past when it came to PM intervals.
> *[VO]* True

Here are some identified causes (IC) and corrective actions (CA)

[IC] Lubrication PM scheduled interval for fan bearing was longer than recommended in lubrication standard.

> *[CA]* Change PM scheduled interval back to recommended lubrication standard

[IC] Production pressures forced a trade-off decision to be made

> *[CA]* Review current procedures and prevent unilateral ability to change SOPs.

[IC] LTA management oversight

> *[CA]* Review the current process to determine whether Management input is required. If not, update.

[IC] Past experience. Reducing downtime for PMs was praised

> *[CA]* Increase emphasis on referencing standards and enforce management of change process.

This concludes our case study and discussion about how to analyze lubrication degradation failures using a holistic approach for root cause analysis. While this case may not have involved a particular lubricant degradation mechanism, we wanted to show the applicability of the process to one of the most common methods of lubrication failures. We hope that these key principles will guide your reliability efforts by reducing the need to react, allowing you time to focus on proactive measures to these foreseeable events. It is always desirable to analyze a high risk than to wait for its costly consequences to crop up unexpectedly.

Remember, we share the same challenge in reliability to defeat this paradigm:

"We NEVER seem to have the time and budget to do things right, but we ALWAYS seem to have the time and budget to do them again!"

We now, collectively, know how to do it right the first time !

References

Ameye, Jo, Dave Wooton, and Greg Livingstone. 2015. *Antioxidant Monitoring as Part of a Lubricant Diagnostics – A Luxury or Necessity.* Rosenheim, Germany. February, 2015.

API (American Petroleum Institute). 2015. API Base Oil Interchangeability Guidelines for Passenger Car Motor Oils and Diesel Engine Oils. Annex E. United States America: API.

Bharat Bhushan. 2013. *Introduction to Tribology, Second Edition.* John Wiley & Sons, Ltd.

Chronic Failure Calculator. Reliability Center, Inc. Accessed March 21, 2021 https://www.reliability.com

Fitch, Jim. 2021. "Root Cause Analysis for lubrication failures." Accessed February 10, 2021. https://www.machinerylubrication.com/Read/857/root-cause?utm_content=153778563&utm_medium=social&utm_source=linkedin&hss_channel=lcp-1539344

Latino, Robert J. PROACT Approach to Healthcare Workshop. January 2005. www.-proactforhealthcare.com

Latino, Robert J. 2019. "How Can v Why: What's the Difference?". Accessed March 21, 2021. https://www.linkedin.com/pulse/how-can-v-why-whats-difference-robert-bob-latino/

Latino, Robert J. 2019. "Is System's Thinking Critical to Root Cause Analysis' (RCA) Success?". Accessed March 21, 2021. https://www.linkedin.com/pulse/systems-thinking-critical-root-cause-analysiss-rca-success-latino/

Latino, Robert J. 2021. "Root Cause vs Shallow Cause Analysis: What's the Difference?". Accessed March 21, 2021. https://www.linkedin.com/pulse/root-cause-vs-shallow-analysis-whats-difference-robert-bob-latino/

Latino, Robert, Kenneth C. Latino, Mark A. Latino. 2019. *Root Cause Analysis: Improving Performance for Bottom-Line Results (5th ed).* CRC Press. Boca Raton, FL.

Leveson, Nancy and Sidney Dekker. 2014. "Get to the Root of Accidents. Systems thinking can provide insights on underlying issues not just their symptoms." Accessed March 21, 2021. https://www.chemicalprocessing.com/articles/2014/get-to-the-root-of-accidents

Mobley, R. Keith. 2020. "Random Failures: Myth or Reality". Accessed March 23, 2021. https://www.linkedin.com/pulse/random-failures-myth-reality-r-keith-mobley/?trackingId=zH4agqgybTLHnYMQhWLKHg%3D%3D

Mortier, Roy M., Malcom F. Fox, and Stefan Orszulik. 2010. *Chemistry and Technology of Lubricants 3rd Edition.* Netherlands: Springer Netherlands.

Noria Corporation. 2021. "How to detect Electrostatic Discharges in Oil." Accessed April 20, 2021. https://www.machinerylubrication.com/Read/30242/detecting-electrostatic-discharges

Pirro, D.M. and A.A. Wessol. 2001. *Lubrication Fundamentals Second Edition, Revised and Expanded.*

PROACT® is a registered trademark of Reliability Center, Inc. Used with permission.

Root Cause Analysis Guidance Document. Washington, DC: US Department of Energy; 1992. DOE publication. DOE-NE-STD-1004-92. Accessed April 29, 2021. http://www.hss.energy.gov/NuclearSafety/techstds/standard/nst1004/nst1004.pdf

Schrank, Katharina, Hubertus Murrenhoff and Christian Stammen. *Investigation of different methods to measure the entrained air content in hydraulic oils.* 2014. *Proceedings of the ASME/BATH 2014 Symposium on Fluid Power & Motion Control FPMC2014*, September 10–12, *Bath, United Kingdom.* Accessed April 18, 2021.

Wang, Yansong and Q. Jane Wang. 2013. Encyclopedia of Tribology, Stribeck curves. United States of America: Springer Boston MA.

Index

Printed in the United States
by Baker & Taylor Publisher Services

Printed in the United States
by Baker & Taylor Publisher Services